Lecture Notes in Mathematics 1484

Hans Delfs

Homology of Locally Semialgebraic Spaces

Springer-Verlag
Berlin Heidelberg New York
London Paris Tokyo
Hong Kong Barcelona
Budapest

Author

Hans Delfs
Fakultät für Mathematik
Universität Regensburg
Universitätsstraße 31
W-8400 Regensburg, FRG

Mathematics Subject Classification (1991): 14G30, 14F45, 55N30, 55N35, 14C17, 54D18

ISBN 3-540-54615-4 Springer-Verlag Berlin Heidelberg New York
ISBN 0-387-54615-4 Springer-Verlag New York Berlin Heidelberg

Printed in Germany

Typesetting: Camera ready by author
Printing and binding: Druckhaus Beltz, Hemsbach/Bergstr.
46/3140-543210 - Printed on acid-free paper

Introduction

The basic task in real algebraic (or better semialgebraic) geometry is to study the set of solutions of a finite system of polynomial inequalities over a real closed field R. Such a set is called a semialgebraic set over R. The semialgebraic subsets of R^n are obtained from the basic open sets

$$U(P) = \{x \in R^n \mid P(x) > 0\},$$

where $P \in R[X_1, \dots, X_n]$ is a polynomial, by finitely many applications of the set theoretic operations of uniting, intersecting and complementing. The interval topology of R induces a topology on every semialgebraic set over R called strong topology. Unfortunately the arising topological spaces are totally disconnected, except in the case $R = \mathbb{R}$. This pathology can be remedied. Strong topology is replaced by semialgebraic topology, a topology in the sense of Grothendieck (cf. [A]): Only open semialgebraic subsets are admitted as „open sets", and essentially only coverings by finitely many open semialgebraic subsets are admitted as „open coverings". These restricted topological spaces are the basic objects studied in semialgebraic topology. (For details concerning this and other notions from semialgebraic geometry we refer to [Br], [BCR], [DK], [DK$_3$], [DK$_4$]).
It is easier to study not only semialgebraic sets which are embedded in some algebraic variety over R but to study more generally spaces which, locally, look like a semialgebraic set. This observation leads to the notion of semialgebraic spaces.
An affine semialgebraic space over R is a ringed space which is isomorphic to a semialgebraic subset N of some affine R-variety V equipped with its sheaf $\mathcal{O}_N$ of semialgebraic functions (cf. [DK]). N is considered in its semialgebraic topology. A locally semialgebraic space $(M, \mathcal{O}_M)$ over a real closed field R is a ringed space ($\mathcal{O}_M$ the structure sheaf) which posesses an admissible open covering $(U_i \mid i \in I)$ such that $(U_i, \mathcal{O}_M \mid U_i)$ is an affine semialgebraic space over R for every $i \in I$ (cf. [DK$_3$], [DK$_4$]).
The category of locally semialgebraic spaces seems to be the appropriate framework for topological considerations over an arbitrary real closed field. (For the study of many questions especially in homotoy theory it turns out to be very useful to work in the even larger category of weakly semialgebraic spaces, cf. [K$_2$]).

This book is a contribution to the fundamentals of semialgebraic topology and consists of two main parts. The first is primarily concerned with the study of sheaves and their cohomology on locally semialgebraic spaces. In the second part we develop a homology theory for locally complete locally semialgebraic spaces over a real closed field R. It is the semialgebraic analogue of the homology for locally compact topological spaces introduced by Borel and Moore ([BM]). Finally we apply the sheaf and homology theory to varieties over an algebraically closed field of characteristic zero and develop a semialgebraic („topological") approach to intersection theory on these varieties.
A first observation is that a sheaf $\mathcal{F}$ on a locally semialgebraic space M over R is not determined by its stalks $\mathcal{F}_x, x \in M$, (Example I.1.7). But fortunately we may embed M into a space $\tilde{M}$ which

- is a topological space in the usual sense
- has the same sheaf theory as M.

If M is a semialgebraic subset of an affine R-variety V, then $\tilde{M}$ is just the associated constructible subset of the real spectrum Sper $R[V]$ of the coordinate ring of V (cf. I, §1). The space $\tilde{M}$ can be endowed with a structure sheaf $\mathcal{O}_{\tilde{M}}$. In this way it becomes a locally ringed space which locally looks like a constructible subset K of some real spectrum Sper A equipped with the sheaf of abstract semialgebraic functions on K defined by Brumfiel ([Br_1]) and Schwartz ([S], [S_1]) (see also [D_2, §1]). Such a space is called an abstract locally semialgebraic space. In Chapter II we investigate sheaves on these spaces. This has the great advantage that abstract locally semialgebraic spaces are topological spaces in the classical sense and therefore classical topological sheaf theory applies to them. (But notice that these spaces are almost never Hausdorff). Since the sheaf theories on M and $\tilde{M}$ coincide, our results are also valid for locally semialgebraic spaces over a real closed field.

In Chapter I we present a short introduction to the theory of abstract locally semialgebraic spaces. We give the definition and explain the connection with the geometric case (§1). Important classes of subsets of a given space are endowed with a subspace structure in §2. Some basic notions which turn out to be useful in sheaf theory are defined and discussed in §3. The subspace $X^{\max}$ of closed points of an abstract locally semialgebraic space X often is a „nice" topologcial space. In §4 we explain, for example, under what conditions $X^{\max}$ is a locally compact space. The most important class of locally semialgebraic spaces consisting of the regular and paracompact spaces is investigated in §5.

Slightly more generally N. Schwartz considers in [S_1] spaces which locally are proconstructible subsets of a variety. He calls these spaces „real closed spaces". Most of the results in chapter II may be easily extended to this more general case. In chapters III and IV we study only the geometric case, locally semialgebraic spaces over a real closed field, and for this purpose the notion of abstract locally semialgebraic spaces as introduced in chapter I is sufficient.

Chapter II is devoted to the study of sheaves on abstract locally semialgebraic spaces. In particular we deal with the cohomology groups of a space X with coefficients in a sheaf $\mathcal{F}$ and support in a family Φ of closed subsets. Usually we are only interested in families Φ of supports which are generated by their locally semialgebraic members. We define paracompactifying support families Φ (this notion is different from the corresponding notion in topological sheaf theory!) and discuss some relations between the properties of Φ and $\Phi \cap X^{\max}$ (§1). The homomorphisms induced in cohomology by locally semialgebraic maps are described in §2. There we also prove that homotopic maps induce the same homomorphism in cohomology. Applying this result we see that, in the geometric case, it is sometimes possible to replace certain families of supports by paracompactifying ones.
Since regular and paracompact locally semialgebraic spaces in many ways show a similar behaviour as paracompact topological spaces, sections of a sheaf over a partially quasicompact subset can be extended to a neighbourhood (§3). This is one of the main reasons why soft sheaves are acyclic as is shown in §4. There is another important class of acyclic sheaves consisting of those sheaves which are flabby in the semialgebraic sense. The results on acyclic sheaves are applied in §5 to describe the cohomology of certain subsets. In particular we learn how the cohomology groups of X and $X^{\max}$ and how the cohomology groups of M and M_{top} are related (where M is a space over the field $\mathbb{R}$ of real numbers and M_{top} is the set M considered in its strong topology).

One of the central results of the book is proven in §6: The cohomology groups of a locally semialgebraic space over a real closed field R with coefficients in a sheaf and arbitrary supports do not change when the base field R is extended. We need this result in §7 to derive the semialgebraic proper base change theorem. This base change theorem is generalized to the case of non proper locally semialgebraic maps in §8. In §9 and §10 we state some facts about the cohomological dimension of geometric spaces and hypercohomology which are needed later on.

Chapter III deals with semialgebraic Borel-Moore-homology. This is a homology theory designed for locally semialgebraic spaces over a real closed field R which are locally complete. Recall that topological Borel-Moore-homology is defined for locally compact spaces and the notion „locally complete" is the semialgebraic substitute for „locally compact". Every affine semialgebraic space M (and more generally every paracompact regular locally semialgebraic space) can be triangulated, i.e. it is isomorphic to a simplicial complex X over R (cf. [DK$_1$, §2], [DK$_3$, II, §4], introduction of chap. III). But in general the simplicial complex X is not closed, i.e. there may be open simplicies σ in X whose faces do not belong to X. Nevertheless these simplicies should contribute to the homology of M. So it seems to be quite natural to take the open simplices as building blocks of a homology theory (and not the closed simplices as in classical simplicial homology). Following this idea we define Borel-Moore-homology groups with constant coefficients and closed supports by use of triangulations and open simplices (§1 and §2). The basic properties of these groups are derived by easy „simplicial arguments". No sheaf theory is needed for this elementary introduction to Borel-Moore-homology. We already sketched this approach in [D$_1$]. These elementary methods suffice to prove the substantial result that every algebraic variety over R posesses a fundamental class (§3).
But of course there are also problems which are difficult to handle by more or less combinatorial considerations. A typical example is Poincaré duality for arbitrary families of supports. It is possible to give an elementary proof but it is long and difficult because the combinatorics involved is rather complicated. Here the use of sheaves turns out to simplify the problem considerably.
Therefore we generalize our definitions in §2 and introduce Borel-Moore-homology groups with arbitrary supports and coefficients in an arbitrary sheaf in §5. In particular the important case of locally constant sheaves is included. The groups are defined by means of a complex $(\Delta_k \mid k \in \mathbb{Z})$ of sheaves of simplical chains (see §4). The sheaves Δ_k are similar to the sheaves of PL-chains in PL-topology. The main differences to PL-theory are that we work with open instead of closed simplices and that a simplicial chain is identified with all its semialgebraic and not only with its linear subdivisions. Semialgebraic subdivision is discussed in §4.
Using weakly semialgebraic spaces and a lot of homotopy theory M. Knebusch was able to prove the really nontrivial result that the semialgebraic homology with compact support may be calculated by singular chains, even if the base field R is non archimedian (cf. [K$_2$]). So the sheaf theoretical approach to homology might also be based, as in classical topology [B], on the sheaves of locally finite singular chains. The semialgebraic triangulation theorem implies that the sheaves of simplicial chains have very nice properties. Therefore it is really an advantage to work with them. For example, the sheaves Δ_k are flabby in the semialgebraic sense and the basic properties of Borel-Moore-homology can be easily deduced from the sheaf theory developed in Chapter II.

In §6 we study the functorial behaviour and in §7 the homotopy invariance of Borel-Moore-homology. A cap product between cohomology and homology is introduced in §8. The proof of a general Poincaré duality theorem is given in §9. In §10 we investigate the behaviour of the Borel-Moore-homology when the real closed groundfield R is enlarged. The relation between the semialgebraic homology of a locally semialgebraic space M over R and the topological Borel-Moore-homology of $\tilde{M}^{\max}$ (resp. M_{top} if $R = \mathbb{R}$) is discussed in §11. Finally we show in §12 that the semialgebraic Borel-Moore-homology groups could also be defined by use of injective resolutions and duals of complexes of sheaves. Such a definition would be analogous to the definition given by Borel and Moore in the topological case ([BM]). (Dualising complexes were also used to define étale Borel-Moore-homology for varieties over arbitrary algebraically closed fields, cf. [DV, exp. VIII]).

In Chapter IV we discuss some aspects of intersection theory. We take an algebraically closed field C of characteristic zero and choose a real closed field $R \subset C$ with $C = R(\sqrt{-1})$. In §1 we prove that every locally isoalgebraic space over C (cf. [Hu] or Appendix, §1, for the definition) posesses a fundamental class. Using Poincaré duality we establish an intersection product in the homology of locally semialgebraic manifolds (§2). This intersection product enables us to define the intersection of locally isoalgebraic cycles on a locally isoalgebraic manifold (§3). In particular we are able to describe the algebraic intersection multiplicities of subvarieties of a smooth algebraic variety V over C in a purely „topologcial" (i.e. semialgebraic) way, and we obtain a multiplicative homomorphism $A_*(V) \to H_*(V)$ from the Chow ring of V to the semialgebraic Borel-Moore-homology.

Throughout this book R is a real closed field. All rings are assumed to be commutative with 1. The real spectrum of a ring A is denoted by Sper A. (The basic theory concerning real spectra is contained in [BCR],[CR] or [L]). The sections of a sheaf $\mathcal{F}$ over an open set U are denoted by $\Gamma(U, \mathcal{F})$. If objects B and C are canonically isomorphic, then we often simply write $B = C$. A list of symbols and an index may be found at the end of this book.

I wish to thank M. Knebusch, R. Huber, R. Robson and N. Schwartz for many helpful discussions on the subject. Finally special thanks are due to M. Richter and R. Bonn for their patience and excellence in typing the manuscript.

Regensburg, May 1991

Hans Delfs

TABLE OF CONTENTS

CHAPTER I: Abstract locally semialgebraic spaces.

This chapter contains a short introduction to the theory of abstract locally semialgebraic spaces. We present the basic definitions and some observations needed in the subsequent parts of these notes.

§1 - Abstract and geometric spaces

We refer the reader to [DK$_3$] for an extensive treatment of locally semialgebraic spaces over the real closed field R. A survey about the basic theory over R is also given in [DK$_4$]. As in algebraic geometry it is sometimes useful to study not only these „geometric spaces" but also a more general class of „abstract spaces". This was made possible by the introduction of abstract semialgebraic functions (cf. [Br$_1$, §3], [S], [S$_1$], [D$_2$, §1]) on the real spectrum (cf. [BCR], [K$_1$]) of an arbitrary ring A (commutative, with 1).

Definition 1. a) Let Sper A be the real spectrum of a ring A. A pair $(K, \mathcal{O}_K)$ consisting of a constructible subset K of Sper A and the sheaf $\mathcal{O}_K$ of abstract semialgebraic functions on K ([D$_2$, §1]) is called a *semialgebraic subspace* of Sper A.
b) An *abstract affine semialgebraic space* is a locally ringed space $(X, \mathcal{O}_X)$ which is isomorphic, as a locally ringed space, to a semialgebraic subspace $(K, \mathcal{O}_K)$ of an affine real spectrum Sper A.
c) An *abstract locally semialgebraic space* is a locally ringed space $(X, \mathcal{O}_X)$ which has an open covering $(X_i \mid i \in I)$ such that $(X_i, \mathcal{O}_X \mid X_i)$ is an abstract affine semialgebraic space for every $i \in I$. Such a covering $(X_i \mid i \in I)$ is called an open affine covering of X. If X is in addition quasicompact, then $(X, \mathcal{O}_X)$ is called an *abstract semialgebraic space*. (Note that „quasicompact" means that X admits a *finite* open affine covering).
d) A *locally semialgebraic map* between abstract locally semialgebraic spaces $(X, \mathcal{O}_X)$ and $(Y, \mathcal{O}_Y)$ is a morphism (f, δ) in the category of locally ringed spaces, i.e. $f: X \to Y$ is a continuous map and $\delta: \mathcal{O}_Y \to f_*\mathcal{O}_X$ is a morphism of sheaves of local rings. We often omit δ in our notation and simply write $f: X \to Y$.

Abstract locally semialgebraic spaces were first defined and studied by N. Schwartz ([S], [S$_1$]). Slightly more generally he considers in [S$_1$] spaces which locally are proconstructible subsets of a variety. He calls these spaces „real closed spaces". We will use many of his definitions and results.

We make the following *general assumption*: All abstract locally semialgebraic spaces $(X, \mathcal{O}_X)$ are assumed to be *quasiseparated* ([S$_1$], II.4.14). This means that the intersection $U \cap V$ of any two open quasicompact subsets U, V of X is also quasicompact, or, equivalently, that $X_i \cap X_j$ is quasicompact for any two members X_i, X_j of an open affine cover $(X_i \mid i \in I)$ of X ([S$_1$, II.4.16]).

In the category of abstract locally semialgebraic spaces arbitrary fibre products exist ([S$_1$, II.3.1]). The residue class field $\mathcal{O}_{X,x}/\mathfrak{m}_{X,x}$ of a space X in a point $x \in X$ {$\mathcal{O}_{X,x}$ the stalk of $\mathcal{O}_X$ in x, $\mathfrak{m}_{X,x}$ the maximal ideal of the local ring $\mathcal{O}_{X,x}$} is real closed and will always be denoted by $k(x)$.

If X is a semialgebraic subspace of Sper A, then $k(x)$ is the real closure of the residue class field $A(\mathfrak{p}(x))$ of A in $\mathfrak{p}(x)$, the support of x, with respect to the ordering of $A(\mathfrak{p}(x))$ which is induced by x. The image of an element $f \in \Gamma(X, \mathcal{O}_X)$ in $k(x)$ under the natural map

$\Gamma(X,\mathcal{O}_X) \to \mathcal{O}_{X,x} \to k(x)$ is denoted by $f(x)$. The elements $f \in \Gamma(X,\mathcal{O}_X)$ are called *locally semialgebraic functions* on X.

Definition 2. Let $(X,\mathcal{O}_X)$ be an affine abstract semialgebraic space. A subset Y of X is called *semialgebraic* (or *constructible*) if Y is a finite union of sets of the form

$$\{x \in X \mid f(x) = 0,\ g_j(x) > 0,\ j = 1,\ldots,s\},$$

$f \in \Gamma(X,\mathcal{O}_X), g_j \in \Gamma(X,\mathcal{O}_X)(1 \le j \le r)$. The set of semialgebraic subsets of X is denoted by $\gamma(X)$.

Of course, if $(X,\mathcal{O}_X)$ is a semialgebraic subspace of Sper A, then $\Gamma(X)$ consists of those constructible subsets of Sper A that are contained in X.

Definition 3. Let $(X,\mathcal{O}_X)$ be an abstract locally semialgebraic space and $(X_i \mid i \in I)$ be an open affine covering of X. A subset Y of X is called *locally semialgebraic* (or *locally constructible*) if $Y \cap X_i$ is a semialgebraic subset of X_i for every $i \in I$. If Y is also quasicompact, then it is called *semialgebraic* (or *constructible*). Since all spaces are assumed to be quasi-separated, the intersection $X_i \cap X_j$ is a semialgebraic subset of X_i, and it is easy to see that our Definition 3 does not depend on the choice of the open affine covering of X. The family of locally semialgebraic (semialgebraic) subsets of X is denoted by $\mathcal{T}(X)$ (by $\gamma(X)$). By $\mathring{\mathcal{T}}(X)$ $\{\mathring{\gamma}(X)\}$ we denote the family of those sets in $\mathcal{T}(X)$ $\{$ in $\gamma(X)\}$ which are open in X, and by $\bar{\mathcal{T}}(X)$ $\{\bar{\gamma}(X)\}$ the family of those sets which are closed in X.

From now on we usually omit the structure sheaves in our notation and simply write X instead of $(X,\mathcal{O}_X)$.

Definition 4. A locally semialgebraic map $f\colon X \to Y$ is called *semialgebraic* (or *quasicompact*) if $f^{-1}(A)$ is a semialgebraic (or, equivalently, a quasicompact) subset of X for every semialgebraic subset $A \in \gamma(Y)$ of Y.

Every locally semialgebraic map $f\colon X \to Y$ whose domain X is a semialgebraic space is semialgebraic.

The concept of an abstract locally semialgebraic space is the natural generalization of the notion of a locally semialgebraic space over a real closed field R as we will explain now.

Example 1.1. Let V be an affine algebraic variety over R. We consider a semialgebraic subspace $(M,\mathcal{O}_M)$ of V (cf. [DK, §7]). By definition, M is a semialgebraic subset of the set $V(R)$ of R-rational points of V, and $\mathcal{O}_M$ is the sheaf of semialgebraic functions on M. Here the set M is considered in its semialgebraic topology (loc. cit.), and a function $f\colon M \to R$ is called semialgebraic if it has a semialgebraic graph and is continuous with respect to the strong topologies. Now let Sper $R[V]$ be the real spectrum of the coordinate ring $R[V]$ of V. The set $V(R)$ is contained in Sper $R[V]$. We may associate a constructible subset $\tilde{A}$ of Sper $R[V]$ to every semialgebraic subset A of $V(R)$ (cf. [CR, §5]). $\tilde{A}$ is defined by the same equalities and inequalities as A. It is the unique constructible subset $\tilde{A}$ of Sper $R[V]$ with $\tilde{A} \cap V(R) = A$. The important point is that the sheaf theories on the semialgebraic space M over R and the topological space $\tilde{M}$ coincide (loc. cit.). More precisely, if $\mathcal{F}$ is a sheaf on M, then we obtain a sheaf $\tilde{\mathcal{F}}$ on $\tilde{M}$ by defining

$$\Gamma(\tilde{U},\tilde{\mathcal{F}}) := \Gamma(U,\mathcal{F})$$

for every open semialgebraic subset U of M. (Note that the sets $\tilde{U}$ form a basis of the topology of $\tilde{M}$). On the other hand, if $\mathcal{G}$ is a sheaf on $\tilde{M}$, then we get a sheaf $\mathcal{F}$ on M by restriction:

$$\Gamma(U,\mathcal{F}) := \Gamma(\tilde{U},\mathcal{G}).$$

Obviously we have $\tilde{\mathcal{F}} = \mathcal{G}$. The sheaf $\mathcal{O}_M$ of semialgebraic functions on M corresponds to the sheaf $\mathcal{O}_{\tilde{M}}$ of abstract semialgebraic functions on $\tilde{M}$. i.e. $\tilde{\mathcal{O}}_M = \mathcal{O}_{\tilde{M}}$ (cf. [D$_2$, 1.9]). We see that it is quite natural to assign the abstract affine semialgebraic space $(\tilde{M},\mathcal{O}_{\tilde{M}})$ to $(M,\mathcal{O}_M)$.

The topological space $\tilde{M}$ has the following useful description. We denote the family of semialgebraic subsets of M by $\gamma(M)$ and the family of open semialgebraic subsets by $\mathring{\gamma}(M)$. Let $Y(M)$ be the set of ultrafilters of the Boolean lattice $\gamma(M)$. The sets

$$Y(U)^M := \{F \in Y(M) \mid U \in F\}, \quad U \in \mathring{\gamma}(M),$$

are the basis of a topology on $Y(M)$. The set $Y(M)$ endowed with this topology is canonically homeomorphic to $\tilde{M}$ ([Brö, p. 260], [CC, §1]).

Now suppose W is another affine R-variety, $(N,\mathcal{O}_N)$ is a semialgebraic subspace of W and $f: M \to N$ is a semialgebraic map (i.e. f is continuous and has a semialgebraic graph). The map f induces a continuous map $\tilde{f}: \tilde{M} \to \tilde{N}$ which may be described as follows: The image $\tilde{f}(F)$ of an ultrafilter $F \in Y(M)$ is the ultrafilter of $\gamma(N)$ generated by the sets $f(A), A \in F$.

Composition with f yields a map of sheaves

$$\delta : \mathcal{O}_N \to f_*\mathcal{O}_M$$

(cf. [DK, §7]). Now observe that $(f_*\mathcal{O}_M)^\sim = \tilde{f}_*\tilde{\mathcal{O}}_M = \tilde{f}_*\mathcal{O}_{\tilde{M}}$. Therefore δ gives a map $\tilde{\delta}: \mathcal{O}_{\tilde{N}} \to \tilde{f}_*\mathcal{O}_{\tilde{M}}$. This map $\tilde{\delta}$ is a map of sheaves of local rings. So our given semialgebraic map $f: M \to N$ yields a morphism $(\tilde{f},\tilde{\delta}) : (\tilde{M},\mathcal{O}_{\tilde{M}}) \to (\tilde{N},\mathcal{O}_{\tilde{N}})$ in the category of abstract semialgebraic spaces.

Using the definitions of locally semialgebraic spaces and locally semialgebraic maps over R in [DK$_3$, I, §1] and Definition 1 above, it is now easy to see that the assignments

$$(M,\mathcal{O}_M) \mapsto (\tilde{M},\mathcal{O}_{\tilde{M}})$$
$$f \mapsto (\tilde{f},\tilde{\delta})$$

extend to a functor from the category of locally semialgebraic spaces over R to the category of abstract locally semialgebraic spaces over R. This functor will always be denoted by $^\sim$. By use of $^\sim$ we may consider the category of locally semialgebraic spaces over R as a *full* subcategory of the category of abstract locally semialgebraic spaces over Sper R. The functor $^\sim$ also preserves fibre products ([S$_1$, III.2.1]). We may and will consider a locally semialgebraic space M over R as a subset of $\tilde{M}$. The set M is dense in $\tilde{M}$ and consists of the points $x \in \tilde{M}$ with $k(x) = R$.

Let M be a locally semialgebraic space over R. As in [DK$_3$] we denote the family of locally semialgebraic subsets (open, closed locally semialgebraic subsets) of M by $\mathcal{T}(M)(\mathring{\mathcal{T}}(M),\bar{\mathcal{T}}(M))$ and the family of semialgebraic subsets (open, closed semialgebraic subsets) by $\gamma(M)(\mathring{\gamma}(M),\bar{\gamma}(M))$.

Proposition 1.2. Intersection with $M, Y \mapsto Y \cap M$, yields canonical bijections

$$\mathcal{T}(\tilde{M}) \xrightarrow{\sim} \mathcal{T}(M), \mathring{\mathcal{T}}(\tilde{M}) \xrightarrow{\sim} \mathring{\mathcal{T}}(M), \bar{\mathcal{T}}(\tilde{M}) \xrightarrow{\sim} \bar{\mathcal{T}}(M),$$
$$\gamma(\tilde{M}) \xrightarrow{\sim} \gamma(M), \mathring{\gamma}(\tilde{M}) \xrightarrow{\sim} \mathring{\gamma}(M), \bar{\gamma}(\tilde{M}) \xrightarrow{\sim} \bar{\gamma}(M).$$

Proof. It suffices to prove this for a semialgebraic subspace M of an affine variety V over R. In this case the result is well known ([CR, §5]).

If $A \in \mathcal{T}(M)$, then $\tilde{A}$ always denotes the (unique) locally semialgebraic subset of $\tilde{M}$ with $\tilde{A} \cap M = A$.

Remark 1.3. Let $f: M \to N$ be a locally semialgebraic map between locally semialgebraic spaces M, N over R. The explicit description of the map $\tilde{f}: \tilde{M} \to \tilde{N}$ in the affine case (cf. Example 1.1) shows that $\tilde{f}^{-1}(\tilde{A}) = \widetilde{f^{-1}(A)}$ for every $A \in \mathcal{T}(N)$. If f is semialgebraic then $f(B) \in \mathcal{T}(N)$ for every $B \in \mathcal{T}(M)$ ([DK$_3$, I.5.3]) and we have $\tilde{f}(\tilde{B}) = \widetilde{f(B)}$.

Again let M be a locally semialgebraic space over R. As in the affine case (cf. Example 1.1) we assign a sheaf $\tilde{\mathcal{F}}$ on $\tilde{M}$ to every sheaf $\mathcal{F}$ on M:

$$\Gamma(Y, \tilde{\mathcal{F}}) := \Gamma(Y \cap M, \mathcal{F}) \quad (Y \in \mathring{\mathcal{T}}(\tilde{M})).$$

Conversely, if $\mathcal{G}$ is a sheaf on $\tilde{M}$ then we obtain a sheaf $\mathcal{F}$ on M by defining

$$\Gamma(U, \mathcal{F}) := \Gamma(\tilde{U}, \mathcal{G}) \quad (U \in \mathring{\mathcal{T}}(M)).$$

Obviously these assignements are inverse to each other. Thus we have

Proposition 1.4. The categories of sheaves on $\tilde{M}$ and M are canonically isomorphic.

So, for all questions concerning sheaves, we may equally well work on $\tilde{M}$ instead on M. For example, since $\Gamma(M, \mathcal{F}) = \Gamma(\tilde{M}, \tilde{\mathcal{F}})$ for every sheaf $\mathcal{F}$ on M the sheaf cohomology theories of M and $\tilde{M}$ coincide.

Corollary 1.5. $H^q(M, \mathcal{F}) = H^q(\tilde{M}, \tilde{\mathcal{F}})$ for every (abelian) sheaf $\mathcal{F}$ on M and every $q \geq 0$.

More generally, we may consider the direct image $f_*\mathcal{F}$ of a sheaf $\mathcal{F}$ on M under a locally semialgebraic map $f: M \to N$ between locally semialgebraic spaces over R. Since $\widetilde{f^{-1}(U)} = \tilde{f}^{-1}(\tilde{U})$ for every $U \in \mathring{\mathcal{T}}(N)$ (Remark 1.3), we have $\widetilde{f_*\mathcal{F}} = \tilde{f}_*\tilde{\mathcal{F}}$. Together with Prop. 1.4 this implies that the right derived functors $R^q f_*$ and $R^q \tilde{f}_* (q \geq 0)$ of f_* and $\tilde{f}_*$ also „coincide".

Corollary 1.6. $(R^q f_* \mathcal{F})^\sim = R^q \tilde{f}_* \tilde{\mathcal{F}}$ for every (abelian) sheaf $\mathcal{F}$ on M and every $q \geq 0$.

In the following we often do not distinguish between sheaves on M and $\tilde{M}$, i.e. sheaves on M are also regarded as sheaves on $\tilde{M}$ (and vice versa). This turns out to be very useful since $\tilde{M}$ is a topological space in the usual sense. So a sheaf $\mathcal{G}$ on M is determined by its stalks $\mathcal{G}_x (:= \tilde{\mathcal{G}}_x), x \in \tilde{M}$. But, in general, it is not determined by the family $(\mathcal{G}_x)_{x \in M}$ of stalks in the points of M.

Example 1.7. Let $M =]0,1[$ be the unit interval over R, considered as a semialgebraic space over R. Then $\tilde{M}$ is the set of ultrafilters of the Boolean lattice γ (]0,1[). Let $\mathcal{G}$ be the associated sheaf on M of the presheaf

$$]a,b[\mapsto \begin{cases} \mathbb{Z} & \text{if} \quad 0=a<b \\ 0 & \text{else} \end{cases}$$

Let x_0 be the ultrafilter generated by the intervals $]0,\varepsilon[, \varepsilon > 0$. Then $\mathcal{G}_{x_0} = \mathbb{Z}$, but $\mathcal{G}_x = 0$ for every $x \in M$.

We close this section with an example of a semialgebraic map.

Example 1.8. A homomorphism $\varphi: A \to B$ induces a continuous map $f = \operatorname{Sper}\varphi: \operatorname{Sper} B \to \operatorname{Sper} A$ (cf. [CR]). Let $\mathcal{O}_{\operatorname{Sper} A}$ and $\mathcal{O}_{\operatorname{Sper} B}$ be the sheaves of abstract semialgebraic functions on $\operatorname{Sper} A$ and $\operatorname{Sper} B$. Composing with f yields a homomorphism $\mathcal{O}_{\operatorname{Sper} A} \to f_*\mathcal{O}_{\operatorname{Sper} B}$ of sheaves of local rings (cf. [D_2, Prop. 1.8]). In this way f becomes a semialgebraic map between the abstract semialgebraic spaces $\operatorname{Sper} B$ and $\operatorname{Sper} A$.

Henceforth an abstract locally semialgebraic space $(X, \mathcal{O}_X)$ will be simply called *space* or *abstract space*. „ *Affine space* " means „affine (abstract) semialgebraic space". A locally semialgebraic space M over R is called a *geometric space* over R. The structure sheaves will usually be omitted in our notation. Unless otherwise stated all maps between spaces are locally semialgebraic maps.

§2 - Subspaces

In this section we explain how certain subsets Y of an abstract space X may be canonically endowed with a subspace structure.

If $Y \subset X$ is open, then obviously $(Y, \mathcal{O}_X \mid Y)$ is an abstract space, called an *open* subspace of X ([S_1, II.2.3]). Note that Y is also quasiseparated.

Our next goal is to define a subspace structure on an arbitrary locally semialgebraic subset Y of X.

Let $U \in \mathring{\gamma}(X)$ be an open affine subspace of X and $f:(U, \mathcal{O}_X \mid U) \xrightarrow{\sim} (U', \mathcal{O}_{U'})$ be an isomorphism onto some semialgebraic subspace $(U', \mathcal{O}_{U'})$ of some real spectrum Sper A. Let $K := U \cap Y$ and $K' := f(K)$. Then K' is a constructible subset of Sper A and we have the sheaf $\mathcal{O}_{K'}$ of abstract semialgebraic functions on $K' \subset$ Sper A. Let $f_1 : K \to K'$ be the restriction of f. We define $\mathcal{O}_K := f_1^* \mathcal{O}_{K'}$. Then $(K, \mathcal{O}_K)$ is an affine (abstract) semialgebraic space, and we have endowed Y with a subspace structure on the open affine part U of X. Now suppose $V \in \mathring{\gamma}(U)$ and $g:(V, \mathcal{O}_X \mid V) \to (V', \mathcal{O}_{V'})$ is an isomorphism onto some semialgebraic subspace $(V', \mathcal{O}_{V'})$ of some real spectrum Sper B. Let $L := V \cap Y, L' := g(L)$ and $g_1 : L \to L'$ be the restriction of g. Let $\mathcal{O}_L$ be the inverse image $g_1^* \mathcal{O}_{L'}$ of the sheaf $\mathcal{O}_{L'}$ of abstract semialgebraic functions on $L' \subset$ Sper B.

Lemma 2.1. $\mathcal{O}_L = \mathcal{O}_K \mid L$.

Proof. cf. [S_1, II.2].

Lemma 2.1. says that the subspace structures $(U \cap Y, \mathcal{O}_{U \cap Y})$ on the affine parts $U \cap Y$, $U \in \mathring{\gamma}(X)$ affine, glue together to form a subspace structure $(Y, \mathcal{O}_Y)$ on Y. Obviously $(Y, \mathcal{O}_Y)$ is an abstract locally semialgebraic space and the inclusion map $Y \hookrightarrow X$ is a locally semialgebraic map. These spaces $(Y, \mathcal{O}_Y)$ are called the *locally semialgebraic subspaces* of $(X, \mathcal{O}_X)$.

Example 2.2. Let $(M, \mathcal{O}_M)$ be a geometric space over R and $N \in \mathcal{T}(M)$. Then we equipped N with a subspace structure $(N, \mathcal{O}_N)$ in [DK_3, §3]. Via the functor $\tilde{}$ the geometric space $(N, \mathcal{O}_N)$ over R corresponds to the subspace $\tilde{N}$ of the abstract space $(\tilde{M}, \mathcal{O}_{\tilde{M}})$ we defined here. This is a trivial consequence of the definitions.

Proposition 2.3. Let $f : X \to Y$ be a locally semialgebraic map between abstract spaces and $Z \in \mathcal{T}(Y)$. Assume $f(X) \subset Z$. Then the map $g : X \to Z$ obtained from f by restriction of the image space is a locally semialgebraic map from X to the subspace Z of Y.

Proof. cf. [S_1, II.2.15].

Finally we consider fibres of locally semialgebraic maps. So let $f : X \to Y$ be a map between abstract spaces and let $y \in Y$. There is a natural locally semialgebraic map

$$\operatorname{Sper} k(y) = \operatorname{Spec} k(y) \to Y$$

mapping the unique point of Spec $k(y)$ to y. We consider the fibre product $X \times_Y \operatorname{Sper} k(y)$. Let $p : X \times_Y \operatorname{Sper} k(y) \to X$ be the projection.

Proposition 2.4. p induces a homeomorphism

$$p_1 : X \times_Y \operatorname{Sper} k(y) \to f^{-1}(y).$$

Proof. cf. [S_1, II.3.2].

We shift the structure sheaf of $X \times_Y \operatorname{Sper} k(y)$ to $f^{-1}(y)$ by p_1. Then $f^{-1}(y)$ becomes an abstract locally semialgebraic space. If $f^{-1}(y)$ is a locally semialgebraic subset of X, then the space structure of $f^{-1}(y)$ defined here coincides with the subspace structure on $f^{-1}(y)$ we considered before ([S_1, II.3.2].

Notation. For an abstract space X we denote the affine space $X \times_{\operatorname{Sper} \mathbb{Z}} \operatorname{Sper} \mathbb{Z}[T_1, \ldots, T_n]$ over X by $\mathbb{A}^n_X$.

Definition 1 (cf. [S_1, II.7.1]). A map $f : X \to Y$ between abstract spaces X and Y is *locally of finite type* if every $x \in X$ has an open neighbourhood $U \in \mathring{\gamma}(X)$ such that the restriction $f \mid U : U \to Y$ admits a factorization

$$\begin{array}{ccc} U & \xrightarrow[\sim]{h} & K \\ & f \searrow \quad \swarrow p & \\ & Y & \end{array}$$

where h is an isomorphism from U onto some semialgebraic subspace K of some affine space $\mathbb{A}^n_y$ and p is induced by the natural projection $\mathbb{A}^n_y \to Y$.

The following result is rather obvious (cf. [S_1, III.1.3]).

Proposition 2.5. Let $f : X \to Y$ be a map between abstract spaces which is locally of finite type. Then the fibre $f^{-1}(y)$ over a point $y \in Y$ is a geometric space over $k(y)$. (Of course this means: There is a geometric space M over $k(y)$ with $\tilde{M} \cong f^{-1}(y)$).

§3 - Some basic notions

We carry over some of the notions introduced in [DK_3, Chap. I] to the abstract case.

Definition 1 ([S_1, II.4.1]). A map $f: X \to Y$ between spaces is called *separated* if the diagonal map $\Delta_f = (\mathrm{id}, \mathrm{id}): X \to X \times_Y X$ is closed.

A geometric space M over R is called separated if it is Hausdorff in its strong topology.

Proposition 3.1. Let M be a geometric space over R. Then M is separated if and only if the map $f: \tilde{M} \to \mathrm{Sper}\, R$ from $\tilde{M}$ to the one-point-space $\mathrm{Sper}\, R$ is separated.

This is Theorem III.3.1 in [S_1].

From now on all geometric spaces are assumed to be separated.

Definition 2 (cf. [DK_3, I, §4, Def. 2]). An abstract space X is called *paracompact* if it possesses a locally finite open affine covering $(X_i \mid i \in I)$.
Here „locally finite" has its usual meaning in topology: Every $x \in X$ has a neighbourhood U which meets only finitely many sets X_i.

Remark 3.2. Since semialgebraic subsets of X are quasicompact, an open covering $(X_i \mid i \in I)$ of X is locally finite if and only if, for every $U \in \mathring{\gamma}(X)$, all but finitely many sets X_i have empty intersection with U.

An immediate consequence of the definition is

Proposition 3.3. Every locally semialgebraic subspace of a paracompact space is paracompact.

Proposition 3.4. An abstract space X is paracompact if and only if every open covering $(X_i \mid i \in I)$ posesses a locally finite refinement $(Y_j \mid j \in J)$ with $Y_j \in \mathring{\gamma}(X)$.

Proof. See [DK_3, I, Prop. 4.5].

N.B. Despite Prop. 3.4 our notation of paracompactness differs from the usual one in topology since abstract spaces are almost never Hausdorff.

Definition 3 (cf. [DK_3, I, §4, Def. 3]). An abstract space X is called *Lindelöf* if it posesses an open covering $(X_i \mid i \in \mathbb{N})$ by countably many semialgebraic subsets $X_i \in \mathring{\gamma}(X)$.

Proposition 3.5. A space X is Lindelöf if and only if every open covering $(X_i \mid i \in I)$ of X has a countable refinement $(Y_j \mid j \in \mathbb{N})$.

Proof. See [DK_3, I, Prop. 4.16].

Proposition 3.6. Let X be a paracompact connected abstract space. Then X is Lindelöf.

Proof. See [DK_3, I, Prop. 4.17].

Counterexample 3.7. If X is paracompact and U is an open subspace of X, then U is in general not paracompact. Consider e.g. $\tilde{M}$ where $M = R$ and R is a non archimedian real closed field which posesses no sequence $(\varepsilon_n \mid n \in \mathbb{N})$ with $\lim_{n\to\infty} \varepsilon_n = 0$. Let $U \subset \tilde{M}$ be the union of the open sets $\widetilde{]a,1[}$ where a runs through all elements $a \in R$ with $0 < a < 1$. Then U has no countable covering by open affine subsets and hence is not paracompact.

Definition 4 (cf. [DK₃, I, §7, Def. 2]). An abstract space X is called *taut* if every $x \in X$ has a closed quasicompact neighbourhood B in X.

Remark 3.8. A space X is taut if and only if the closure $\bar{A}$ of every semialgebraic subset $A \in \gamma(X)$ of X is quasicompact. (N.B. We do not know in general whether $\bar{A} \in \mathcal{T}(X)$). Thus X is taut if and only if the closure $\bar{A}$ of any semialgebraic subset $A \in \gamma(X)$ of X has an open semialgebraic neighbourhood $U \in \mathring{\gamma}(X)$ in X.

Proposition 3.9. If an abstract space X is Lindelöf and taut, then it is paracompact.

Proof. See [DK_3, I, Prop. 4.18].

Proposition 3.10. A paracompact space X is taut.

Proof. See [DK_3, I, Prop. 4.6].

Proposition 3.11. Let M be a geometric space over R. Then M is paracompact (Lindelöf, taut) if and only if $\tilde{M}$ is paracompact (Lindelöf, taut).

This follows immediately from the definitions and the remarks 3.2 and 3.8.

Notation. For points x, y in an abstract space X we write $x \subset y$ if y is a specialization of x, i.e. $y \in \overline{\{x\}}$.

Definition 5 a) ([S_1, II.5.1]). An abstract space X is called *regular* if, for any $x, y, z \in X$ with $x \subset y$ and $x \subset z$, we either have $y \subset z$ or $z \subset y$. In other words: A space X is regular if and only if the specializations of x form a chain for every $x \in X$.
b) An abstract space X is called *maximally closed* if and only if every $x \in X$ specializes to a closed point in X.

The set of closed points of an abstract space X is denoted by $X^{\max}$. More generally, if $A \subset X$ is closed, then we denote $A \cap X^{\max}$ by $A^{\max}$.

Remarks 3.12. a) If X is a regular and maximally closed space then we have a canonical retraction $r_X : X \to X^{\max}$ mapping $x \in X$ to the unique closed point y with $x \subset y$.
b) Affine spaces are regular as is well known. Conversely, if a space X is quasicompact (i.e. semialgebraic) and regular, then it is affine ([S_1, II.5.8]).
c) Semialgebraic spaces, taut spaces and all spaces $\tilde{M}$, M a geometric locally semialgebraic space over some real closed field R, are maximally closed.

We need the following two lemmas.

Lemma 3.13. Let X be an affine space and $(U_i \mid i \in I)$ be a family in $\gamma(X)$. Then $\cap(U_i \mid i \in I)$ is not empty if and only if all finite intersections are not empty.

This fact is well known and follows from the compactness of the constructible topology of a real spectrum $\operatorname{Sper} A$ (cf. [CC], [R]).

Lemma 3.14. Let X be an affine space and A be a quasicompact subset of X. Denote the family $\{U \in \mathring{\gamma}(X) \mid A \subset U\}$ of open semialgebraic neighbourhoods of A in X by $\mathfrak{U}_A$. Then $\cap\mathfrak{U}_A$ consists of those points in X which have a specialization in A.

Proof. If $x \subset y$ and $y \in A$ then obviously $x \in \cap\mathfrak{U}_A$. Conversely let $x \in \cap\mathfrak{U}_A$. Assume that x has no specialization in A. Then every $y \in A$ has an open semialgebraic neighbourhood U_y with $x \notin U_y$. Since A is quasicompact, we then find an element $V \in \mathfrak{U}_A$ with $x \notin V$. This contradiction shows that x has indeed a specialization in A.

Proposition 3.15. Let X be an abstract space. If X is regular and taut then for every $x \in X^{\max}$ and every closed subset A of X with $x \notin A$, there exist neighbourhoods $U, V \in \mathring{T}(X)$ of x, A with $U \cap V = \emptyset$.
The converse is true if X is maximally closed.

Proof. First assume that X is regular and taut. Let A be a closed subset of X and $x \in X^{\max} \setminus A$. Let $U \subset X \setminus A$ be an open semialgebraic neighbourhood of x. There is some $V \in \mathring{\gamma}(X)$ with $\bar{U} \subset V$ (Remark 3.8). Note that V is affine (Remark 3.12.b). Let $\mathfrak{A}$ be the family of neighbourhoods B of x in V with $B \in \bar{\gamma}(V)$. It suffices to find some $B \in \mathfrak{A}$ with $B \subset U$. By Lemma 3.13 it suffices to show that $(\cap\mathfrak{A}) \cap (V \setminus U) = \emptyset$. Suppose there is a point y in this intersection. We may choose y in $V^{\max}$. Obviously $y \neq x$ and y and x cannot be separated by disjoint neighbourhoods. Thus $(\cap\mathfrak{U}_x) \cap (\cap\mathfrak{U}_y) \neq \emptyset$ by Lemma 3.13 where $\mathfrak{U}_x$ and $\mathfrak{U}_y$ are the families of open semialgebraic neighbourhoods of x and y in V. But every point $z \in (\cap\mathfrak{U}_x) \cap (\cap\mathfrak{U}_y)$ is a common generalization of x and y (Lemma 3.14). This is a contradiction.

Now we assume that X is maximally closed and we can separate points in $X^{\max}$ from closed subsets of X by open neighbourhoods in X. If $x \in X^{\max}$ and U is an open semialgebraic neighbourhood of x, then U must contain a closed neighbourhood B of x since x and $X \setminus U$ can be separated. Since U is quasicompact, B is also quasicompact and we see that X is taut. Our hypothesis says in particular that any different points $x, y \in X^{\max}$ can be separated by disjoint neighbourhoods in X. This implies that X is regular. q.e.d.

Recall that a geometric space M over R is defined to be regular if points $x \in M$ and sets $A \in \bar{T}(M)$ can be separated by disjoint open locally semialgebraic neighbourhoods provided $x \notin A$ (cf. [DK_3, I, §4, Def. 1]).

Proposition 3.16. Let M be a regular geometric space over R. Then $\tilde{M}$ is a regular abstract space.

Proof. Let $(M_i \mid i \in I)$ be an admissible covering of M by open semialgebraic subsets. Let x, y, z be points in $\tilde{M}$ with $x \subset y$ and $x \subset z$. We choose indices $i, j \in I$ such that $y \in \tilde{M}_i$ and $z \in \tilde{M}_j$. Of course $x \in \tilde{M}_i \cap \tilde{M}_j$. Since M is regular, $M_i \cup M_j$ is an affine semialgebraic space ([Ro], [DK_3, I, 4.1]). Hence $(M_i \cup M_j)^{\sim} = \tilde{M}_i \cup \tilde{M}_j$ is a regular abstract space (Remark 3.12.b) and we see that either $y \subset z$ or $z \subset y$. q.e.d.

The converse of Prop. 3.16. does not hold in full generality.

Example 3.17. We consider the semialgebraic subspaces

$$M_k :=]\frac{1}{k}, 1] \times [0,1] (k \in \mathbb{N}), M_0 := ([0,1] \times]0,1[) \cup \{(0,0)\}$$

of $\mathbf{R}^2$. Let $N_k := M_0 \cup M_k$ and let $M := \varinjlim N_k$ be the inductive limit of the affine semialgebraic spaces N_k ([DK$_3$, I, §2]). M is a locally semialgebraic space over $\mathbf{R}$ with admissible open covering $(M_k \mid k = 0,1,2,\ldots)$. It is not regular because $x := (0,0)$ and the closed locally semialgebraic subset $A :=]0,1] \times \{0\}$ of M cannot be separated by disjoint neighbourhoods. But $\tilde{M} = \cup(\tilde{M}_k \mid k = 0,1,2,\ldots)$ is a regular abstract space.

The reason why the converse of 3.16 fails in Example 3.17 is that M is not taut. (The closure of M_0 is not semialgebraic).

Proposition 3.18. Let M be a geometric space over R. Assume that $\tilde{M}$ is regular and taut. Then M is regular.

Proof. Follows from Proposition 3.15.

The propositions 3.10, 3.11, 3.16 and 3.18 imply

Corollary 3.19. A geometric space M over R is regular and paracompact if and only if the abstract space $\tilde{M}$ is regular and paracompact.

Example 3.17. provides us with another counterexample.

Example 3.20. If A is a closed locally semialgebraic subset of a geometric space M, then $\{\tilde{U} \mid U \in \mathring{T}(M), A \subset U\}$ is in general no fundamental system of neighbourhoods of $\tilde{A}$ in $\tilde{M}$, even if M is regular.
Consider, for example, the space M and its subset A in Example 3.17. Then A is also a closed locally semialgebraic subset of $N := M \setminus \{(0,0)\}$ and N is regular. Now $W := \{(x,t) \in N \mid x > 0, 0 \leq t < \exp(-\frac{1}{x})\}$ is an open neighbourhood of A in the strong topology which does not contain an open locally semialgebraic neighbourhood of A. Let $\tilde{W}$ be the set of all points $x \in \tilde{M}$ which have a specialization in W. $\tilde{W}$ is an open neighbourhood of $\tilde{A}$ in $\tilde{M}$ and $\tilde{W} \cap M = W$. Hence $\tilde{W}$ does not contain a set $\tilde{U}$ with $U \in \mathring{T}(M), A \subset U$.

§4 - The closed points

The (topological) subspace $(\operatorname{Sper} A)^{\max}$ of an affine real spectrum $\operatorname{Sper} A$ consisting of the closed points is a nice topological space. It is Hausdorff and compact. Here we pose the following question: What conditions do guarantee that the closed points $X^{\max}$ of an abstract space X form a „nice" topological space?
The first result is

Proposition 4.1. Let X be an abstract space. If X is regular then $X^{\max}$ is a Hausdorff space. The converse is true if X is maximally closed and $X^{\max}$ lies dense in X.

N.B. If M is a geometric space over R and $X = \tilde{M}$ then M and hence also $X^{\max}$ is dense in X.

Proof. We consider a regular space X and points $x, y \in X^{\max}, x \neq y$. There is an open semialgebraic neighbourhood $U \in \mathring{\gamma}(X)$ of x and y. By Remark 3.12.b U is affine. Since $U^{\max}$ is Hausdorff and $X^{\max} \cap U \subset U^{\max}$, we can separate the points x and y by disjoint neighbourhoods in $X^{\max}$.
Now suppose that X is maximally closed, $X^{\max}$ is Hausdorff and is a dense subset of X. We claim that X is regular. It suffices to find disjoint open neighbourhoods U, V in X of any two different closed points x, y of X. But there are open neighbourhoods U, V of x, y with $U \cap V \cap X^{\max} = \emptyset$. Since $X^{\max}$ is dense this implies $U \cap V = \emptyset$.

Example 4.2. There are spaces X with $X^{\max}$ a Hausdorff space although X is not regular: Let $\overline{\mathsf{R}(X)}$ be the real closure of $\mathsf{R}(X)$ with respect to the ordering given by the ultrafilter $x_0 \in \operatorname{Sper} \mathsf{R}[X]$ which is generated by the intervals $]0, \varepsilon[, \varepsilon > 0$. Let B be the convex hull of $\mathsf{R}[X]_{(0)}$ in $\overline{\mathsf{R}(X)}$. Then $\operatorname{Sper} B$ can be considered as a subset of $\operatorname{Sper} \mathsf{R}[X]$. It consists of the two points 0 and x_0. Now we take two copies X_1, X_2 of $\operatorname{Sper} B$ and glue them along the open constructible subset $\{x_0\}$ of $\operatorname{Sper} B$. We obtain a space X which obviously is not regular. But $X^{\max}$ is discrete and in particular Hausdorff.

Now we consider a regular and maximally closed space X. Recall from Remark 3.12.a that we have a canonical retraction $r_X : X \to X^{\max}$. For affine spaces X this retraction was studied e.g. in [CC]. It is shown there that r_X is continuous in this case.

Proposition 4.3. The canonical retraction $r_X : X \to X^{\max}$ is continuous if and only if X is taut.

Proof. First assume that X ist taut. Let $x \in X$. Since X is taut we find open neighbourhoods $U, V \in \mathring{\gamma}(X)$ of x with $\bar{U} \subset V$ (Remark 3.8). Being a regular space, V is affine (Remark 3.12.b)). We have $U \cap X^{\max} = U \cap V^{\max}$, hence $r_X \mid U = r_V \mid U$. We know from the affine case that r_V is continuous. Hence r_X is also continuous.

Now we suppose r_X is continuous. Let $x \in X$ and $U \in \mathring{\gamma}(X)$ be a neighbourhood of $r_X(x)$. The set $r_X^{-1}(U \cap X^{\max})$ is open, and it is contained in U. We choose a neighbourhood $V \in \mathring{\gamma}(X)$ of x with $V \subset r_X^{-1}(U \cap X^{\max})$. Since $\bar{V}$ consists of all specializations of points in V (see Lemma 4.4 below), $\bar{V}$ is also contained in $r_X^{-1}(U \cap X^{\max})$ and hence in U. This implies that $\bar{V}$ is quasicompact. We see that X is taut.

The following very useful lemma is well known (cf. [CC, Prop. 1]).

Lemma 4.4: Let X be an abstract space and $Y \in \mathcal{T}(X)$. Then $y \in \bar{Y}$ if and only if y is the specialization of a point $x \in Y$. In particular, Y is closed if and only if it is closed under specializations, and Y is open if and only if it is closed under generalizations.

Proof. Of course all specializations of Y belong to $\bar{Y}$. Let $y \in \bar{Y}$. Let $U \in \mathring{\gamma}(X)$ be an affine semialgebraic neighbourhood of y. Then $U \cap Y \in \gamma(U)$ and $y \in \overline{U \cap Y}$. We denote the family of all open semialgebraic neighbourhoods of y in U by $\mathfrak{U}_y$. From Lemma 3.13 we know that $(\cap \mathfrak{U}_y) \cap Y \neq \emptyset$. Now we conclude from Lemma 3.14 that y has a generalization in Y.

Proposition 4.5. Let X be an abstract space. If X is taut and regular, then $X^{\max}$ is a locally compact space. The converse is true if X is maximally closed and $X^{\max}$ is dense in X.

Proof. Let X be taut and regular. Then $X^{\max}$ is Hausdorff by Proposition 4.1. Consider a point $x \in X^{\max}$. We choose neighbourhoods $U, V \in \mathring{\gamma}(X)$ of x with $\bar{U} \subset V$ (Remark 3.8). Being regular V is an affine space (Remark 3.12.b)). Hence $V^{\max}$ is compact ([Brö, 2.7], [CC, Prop. 2]) and $\bar{U} \cap X^{\max} = \bar{U} \cap V^{\max}$ is a compact neighbourhood of x in $X^{\max}$.
Now we assume that X is maximally closed and that $X^{\max}$ is dense in X and is a locally compact space. This includes that $X^{\max}$ is Hausdorff. Hence X is regular by Proposition 4.1. Let $x \in X^{\max}$. We have to find a neighbourhood $U \in \mathring{\gamma}(X)$ of x such that $\bar{U}$ is quasicompact. Let A be a compact neighbourhood of x in $X^{\max}$ and choose $U \in \mathring{\gamma}(X)$ with $U \cap X^{\max} \subset A$. Since $X^{\max}$ is dense in X and A is closed in $X^{\max}$ we conclude $\bar{U} \cap X^{\max} = \overline{(U \cap X^{\max})} \cap X^{\max} \subset \bar{A} \cap X^{\max} = A$. Hence $\bar{U} \cap X^{\max}$ is compact. This implies that $\bar{U}$ is quasicompact.

Remark 4.6. Let M be a locally complete geometric space over R. We know that M is regular ([DK$_3$, §7]). But $\tilde{M}^{\max}$ is not necessarily a locally compact space. We consider Example 3.15 in ([DK$_3$, Chap. I]. Let $R = \mathbb{R}$ and

$$M_k :=]\tfrac{1}{k}, 1] \times]0,1[\qquad (k \in \mathbb{N}),$$
$$M_0 := \{(x,y) \in [0,1] \times]0,1[\mid x^2 + (y - \tfrac{1}{2})^2 < \tfrac{1}{4}\},$$
$$N_k := M_k \cup M_0, M := \varinjlim N_k.$$

Then M is locally complete, but the ultrafilter $x_0 \in \tilde{M}^{\max}$ indicated in the picture below has no compact neighbourhood in $\tilde{M}^{\max}$.

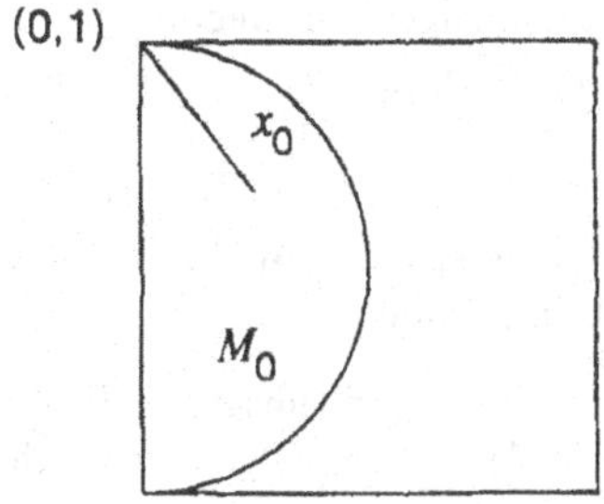

§5 - Regular and paracompact spaces

The most important spaces are the regular and paracompact ones. Some of their properties are described in this section.

Proposition 5.1. Let X be a regular space and $(X_i \mid i \in I)$ be a locally finite covering of X by open semialgebraic subsets $X_i \in \mathring{\gamma}(X)$. (In particular X is paracompact). Then there exist a covering $(U_i \mid i \in I)$ of X by open semialgebraic subsets $U_i \in \mathring{\gamma}(X)$ and a covering $(B_i \mid i \in I)$ of X by closed semialgebraic subsets $B_i \in \bar{\gamma}(X)$ such that $U_i \subset B_i \subset X_i$ for every $i \in I$.

Proof. A proof for affine spaces X may be found in [D_2, 2.1]. In the general case we observe that all semialgebraic subsets $A \in \gamma(X)$, in particular all finite unions $X_{i_0} \cup \ldots \cup X_{i_r}$, are affine spaces since X is regular. Starting from the known affine case, now the same proof as that of [DK_3, I, 4.11] gives the desired result.

Theorem 5.2 (Partition of unity). Let X be a regular and paracompact space and $(X_i \mid i \in I)$ be a locally finite covering of X by open semialgebraic subsets $X_i \in \mathring{\gamma}(X)$. Then there exists a family $(f_i \mid i \in I)$ in $\Gamma(X, \mathcal{O}_X)$ such that $0 \leq f_i(x) \leq 1$ for every $i \in I$ and every $x \in X$, $\mathrm{supp}(f_i) \subset X_i$ for every $i \in I$ and $\sum_{i \in I} f_i(x) = 1$ for every $x \in X$.
Here $\mathrm{supp}(f)$ denotes as usual the closure of the set $\{x \in X \mid f(x) \neq 0\}$.

Proof. Let $(U_i \mid i \in I)$ and $(B_i \mid i \in I)$ be „shrinkings" of $(X_i \mid i \in I)$ as described in Proposition 5.1. By [S_1, I.3.27] we find abstract semialgebraic functions $g_i \in \Gamma(X_i, \mathcal{O}_{X_i})$, $g_i \geq 0$, with $U_i = \{x \in X_i \mid g_i(x) > 0\}$. We can extend g_i by zero to a locally semialgebraic function $h_i \in \Gamma(X, \mathcal{O}_X)$. Defining

$$f_i(x) := \frac{h_i(x)}{\sum_{j \in I} h_j(x)}$$

we obtain the desired family $(f_i \mid i \in I)$.

Theorem 5.3. Let X be a regular and paracompact space and A, B be disjoint closed subsets of X. Then there is a locally semialgebraic function $f \in \Gamma(X, \mathcal{O}_X)$ with $0 \leq f \leq 1$ and $A \subset f^{-1}(0), B \subset f^{-1}(1)$. If $A, B \in \bar{T}(X)$, then we may choose f such that $A = f^{-1}(0), B = f^{-1}(1)$.

Proof. This is known to be true for affine spaces ([S_1, I.3.28]) and thus follows for a regular paracompact space by use of a partition of unity.

Theorem 5.3 says in particular that disjoint closed subsets A, B of a regular and paracompact space can be separated by disjoint open locally semialgebraic neighbourhoods.

Definition 1. Let X be an abstract space. A (topological) subspace Y of X is called *partially quasicompact* if $Y \cap A$ is quasicompact for every $A \in \bar{\gamma}(X)$.

Examples 5.4. a) If $A \in \mathcal{T}(X)$, then A is partially quasicompact.
b) All closed subsets A of X are partially quasicompact.
c) The closed points X^{max} of X form a partially quasicompact subspace of X.
d) Let M be a partially complete geometric space over $\mathbf{R}$ (cf. [DK$_3$, I, §6]). We consider M in its strong topology and denote this topological space by M_{top}. Then M_{top} is a partially quasicompact subspace of $\tilde{M}$. Indeed, every closed semialgebraic subset $A \in \bar{\gamma}(M)$ of M is complete. Hence it is compact in its strong topology ([DK, 9.11]). But $\tilde{A} \cap M_{top} = A_{top}$. Note that $M_{top} = X^{max}$ if M is assumed to be taut. (This is not true in general as Example 4.6 shows).
e) Let $f: X \to Y$ be a locally semialgebraic map between abstract spaces X, Y. Then all fibres $f^{-1}(y)$ are partially quasicompact.

Indeed, for a proof of this fact we may assume that Y is affine semialgebraic. If $y \in Y$ then there is a set $U \in \mathring{\gamma}(Y)$ with $y \in U^{max}$. The fibre $f^{-1}(y)$ is closed in $f^{-1}(U)$ and $f^{-1}(U) \in \mathring{\mathcal{T}}(X)$. Hence $f^{-1}(y)$ is partially quasicompact by a).

Lemma 5.5 (cf. Example 3.20). Let A be a partially quasicompact subset of a space X. Assume that A posesses a regular and paracompact neighbourhood $Y \in \mathcal{T}(X)$. Then the family $\mathfrak{U}_A$ of all open locally semialgebraic neighbourhoods $V \in \mathring{\mathcal{T}}(X)$ of A is a fundamental system of neighbourhoods of A in X.

Proof. Replacing X by Y we may assume that X is paracompact and regular. Let $(X_i \mid i \in I)$ be an open locally finite affine covering of X. We choose „shrinkings" $(U_i \mid i \in I)$ and $(B_i \mid i \in I)$ of $(X_i \mid i \in I)$ as described in Proposition 5.1. Let U be an open neighbourhood of A in X. Since $A \cap B_i$ is quasicompact there are *finitely many* $V_{ij} \in \mathring{\gamma}(X), j \in J(i)$, with $V_{ij} \subset U \cap X_i$ and $A \cap B_i \subset \bigcup(V_{ij} \mid j \in J(i))$ for every $i \in I$. Let $V_i = \bigcup(V_{ij} \mid j \in J(i))$ and $V = \bigcup(V_i \mid i \in I)$. Then V is an element of $\mathcal{T}(X)$ since $(V_i \mid i \in I)$ is a locally finite family. The set V contains A, it is open and it is contained in U.

Remark 5.6. Let A be a partially quasicompact subset of a regular and paracompact space X. Assume that A is closed under specializations. (This condition implies that $r_X(X) \subset A$. It is fulfilled if A is closed in X or $A \subset X^{max}$). Then every open neighbourhood U of A in X contains a closed locally semialgebraic neighbourhood $B \in \bar{\mathcal{T}}(X)$ of A. Thus the closed locally semialgebraic neighbourhoods also form a fundamental system of neighbourhoods of A.

Proof. Let U be an open neighbourhood of A in X. Let $y \in \bar{A}$. Then y lies in the closure of some set $A \cap B_i$ where the sets B_j are chosen as in the preceding proof. From Proposition 3.15 we know that every $x \in A \cap B_i \cap X^{max}$ posesses an open neighbourhood U_x in X whose closure $\bar{U}_x$ is contained in U. Our hypothesis $r_X(A) \subset A$ guarantees that these sets U_x cover $A \cap B_i$. But then already finitely many of the sets U_x cover $A \cap B_i$ and hence $y \in \bar{U}_x \subset U$ for some $x \in A \cap B_i \cap X^{max}$. We see that $\bar{A} \subset U$. The assertion now follows from Theorem 5.3.

Theorem 5.7. Let X be an abstract space. If X is paracompact and regular, then X^{max} is a paracompact topological space (in the sense of topology [Sch, I.8.5]). The converse is true if X is maximally closed and X^{max} is dense in X.

Proof. Let X be regular and paracompact. Then $X^{\max}$ is Hausdorff by Prop. 4.1. Let $(V_i \mid i \in I)$ be an open covering of $X^{\max}$. We choose open subsets U_i of X with $U_i \cap X^{\max} = V_i$. Then $(U_i \mid i \in I)$ is an open covering of X. It posesses a locally finite refinement $(W_j \mid i \in I)$. Now $(W_j \cap X^{\max} \mid j \in J)$ is a locally finite refinement of $(V_i \mid i \in I)$.

Conversely assume that X is maximally closed and $X^{\max}$ is dense in X and $X^{\max}$ is a paracompact topological space. Then X is regular by Prop. 4.1. Let $(U_i \mid i \in I)$ be a locally finite open cover of $X^{\max}$ such that every U_i is contained in an open semialgebraic subset $X_i \in \mathring{\gamma}(X)$. Let $(V_i \mid i \in I)$ be a refinement of $(U_i \mid i \in I)$ with $\bar{V}_i \cap X^{\max} \subset U_i$ ([Sch, I.8.6.2]). Here $\bar{V}_i$ denotes the closure of V_i in X. Since $\bar{V}_i$ is contained in X_i and X_i is quasicompact, $\bar{V}_i \cap X^{\max}$ is a compact set. Therefore we may choose sets $W_i \in \mathring{\gamma}(X)$ with $W_i \subset X_i$ and $\bar{V}_i \cap X^{\max} \subset W_i \cap X^{\max} \subset U_i$. As we just saw $X^{\max}$ is locally compact, hence X is taut (Prop. 4.5). This implies that $\bar{W}_i$ is quasicompact (Remark 3.8) and hence $\bar{W}_i \cap X^{\max}$ is compact. Since $(U_i \mid j \in I)$ is a locally finite cover and $W_j \cap X^{\max} \subset U_j$, we conclude that, for fixed $i, W_i \cap W_j \cap X^{\max}$ and hence $W_i \cap W_j$ is non empty for only finitely many $j \in I$. Thus we have found a locally finite covering $(W_i \mid i \in I)$ of X by open semialgebraic subsets, and X is indeed paracompact.

Remark 5.8. The assumption „ $X^{\max}$ dense in X" is not superfluous as Example 4.2 shows. Take the valuation ring B considered there. Then glue infinitely many copies X_i of Sper B along the open constructible subset $\{x_0\}$ of Sper B. The resulting space X is neither regular nor paracompact although $X^{\max}$ is discrete and hence paracompact.

CHAPTER II: Sheaf theory on locally semialgebraic spaces

We fix a commutative ground ring Λ with 1. The sheaves studied in this chapter are assumed to be sheaves of Λ-modules. Unless otherwise stated all tensor products are taken over Λ. (In some examples we take $\Lambda = \mathbb{Z}$ as the reader will easily realize). To distinguish between those subspaces Y of an abstract space X which are endowed with an induced structure sheaf and are locally semialgebraic spaces themselves and the purely topological subspaces of X, we call the latter ones simply „subsets". Therefore we agree on the following: All subsets Y of an abstract space X are considered as topological spaces equipped with the topology which is induced by the topology of X. (But of course geometric spaces M over R are still considered in their semialgebraic topology and not in their strong topology which is induced by the inclusion $M \subset \tilde{M}$).

We assume that the reader is familiar with classical topological sheaf theory (cf. e.g. [B], [G]). We are mainly interested in the cohomology groups $H^q_\Phi(X, \mathcal{F})$ of a space X with coefficients in some sheaf $\mathcal{F}$ and supports in some family Φ of closed subsets (cf. §1). Note that $H^q_\Phi(X, \mathcal{F})$ may be computed by an arbitrary flabby (cf. §4, Def. 1) resolution $0 \to \mathcal{F} \to \mathcal{J}^\bullet$ of $\mathcal{F}$ by Λ-modules ([B, II.4.1], [G, II.4.7.1]). Considering, for a moment, $\mathcal{F}$ and all $\mathcal{J}^k$ as $\mathbb{Z}$-modules we see that $0 \to \mathcal{F} \to \mathcal{J}^\bullet$ is also a resolution of the $\mathbb{Z}$-module $\mathcal{F}$ by flabby sheaves of $\mathbb{Z}$-modules. Hence $H^q_\Phi(X, \mathcal{F})$ does not depend on the ground ring Λ, it is determined by the structure of the $\mathbb{Z}$-module $\mathcal{F}$. Thus we could have chosen $\Lambda = \mathbb{Z}$ from the beginning. Nevertheless it is sometimes an advantage to work over an arbitrary ground ring.

Most of our definitions are given in the category of abstract spaces. We implicitly assume that all notions are transfered to the category of geometric spaces M over R by the functor $M \mapsto \tilde{M}$ and hence are available there. We illustrate this by two examples.

If G is a Λ-module and X is a space, then G_X is defined to be the *constant sheaf* G on X. It is the associated sheaf of the presheaf which assigns G to every open subset U of X. Now, if M is a geometric space over R, then G_M denotes the sheaf on M which corresponds to $G_{\tilde{M}}$ (cf. I, §1). Obviously G_M is the associated sheaf on M of the presheaf which assigns G to every open locally semialgebraic subset U of M. (More precisely, $\Gamma(U, G_M) = \prod_{\pi_0(U)} G$ where $\pi_0(U)$ is the set of connected components of U).

Definition 1. A sheaf $\mathcal{F}$ on an abstract space X is said to be *locally constant* if every $x \in X$ has an open neighbourhood U such that $\mathcal{F} \mid U$ is constant.

Thus a sheaf $\mathcal{F}$ on a geometric space M is locally constant if there is an admissible covering $(U_i \mid i \in I) \in \mathrm{Cov}(M)$ of M by open locally semialgebraic subsets such that $\mathcal{F} \mid U_i$ is constant for every $i \in I$.

If $f: Y \to X$ is a continuous map between topological spaces (e.g. a locally semialgebraic map between abstract spaces) and $\mathcal{F}$ and $\mathcal{G}$ are sheaves on Y and X, then $f_*\mathcal{F}$ as usual denotes the direct image of $\mathcal{F}$ and $f^*\mathcal{G}$ the inverse image of $\mathcal{G}$ (cf. [G, II.1.12, 1.13]). The higher images of $\mathcal{F}$ are denoted by $R^q f_* \mathcal{F}$. Recall that $R^q f_* \mathcal{F}$ is the associated sheaf on X of the presheaf $U \mapsto H^q(f^{-1}(U), \mathcal{F})$. Now assume that f is an inclusion map. Then $f^*\mathcal{G}$ is simply denoted by $\mathcal{G} \mid Y$. If $s \in \Gamma(X, \mathcal{G})$ then $s \mid Y$ denotes the restriction of s to Y, i.e. $s \mid Y$ is the image of s under the natural map $\Gamma(X, \mathcal{G}) \to \Gamma(Y, \mathcal{G} \mid Y)$.

To simplify our notation we will nearly always write

$\Gamma(Y, \mathcal{G})$ instead of $\Gamma(Y, \mathcal{G} \mid Y)$ and

$H^q_\Phi(Y, \mathcal{G})$ instead of $H^q_\Phi(Y, \mathcal{G} \mid Y)$.

Sometimes we do not distinguish between a sheaf $\mathcal{F}$ on a geometric space M over R and the corresponding sheaf $\tilde{\mathcal{F}}$ on $\tilde{M}$ in our notation and simply write $\mathcal{F}$ instead of $\tilde{\mathcal{F}}$.

§1 - Families of supports

First we recall the definition ([B, I.6.1]). Let X be a topological space.

Definition 1. A *family of supports* is a family Φ of closed subsets of X such that
a) A closed subset of a member of Φ is in Φ.
b) Φ is closed under finite unions.

Let $\mathcal{F}$ be a sheaf on X and $s \in \Gamma(X,\mathcal{F})$ be a section of $\mathcal{F}$. For every $x \in X$ the section s yields an element in the stalk $\mathcal{F}_x$. This element is always denoted by s_x and called the *germ* of s in x.

Definition 2. The *support* $\mid s \mid$ ($=$ supp s) of s is the set of those points $x \in X$ where $s_x \neq 0$.

Note that $\mid s \mid$ is a closed subset of X. For any family Φ of supports on X we define

$$\Gamma_\Phi(X,\mathcal{F}) := \{s \in \Gamma(X,\mathcal{F}) \mid \operatorname{supp} s \in \Phi\}.$$

The functor $\mathcal{F} \mapsto \Gamma_\Phi(X,\mathcal{F})$ is left exact. Its right derived functors $R^q\Gamma_\Phi(X,-)$ yield the cohomology groups with supports in Φ (cf. [B], [G]):

$$H^q_\Phi(X,\mathcal{F}) = R^q\Gamma_\Phi(X,\mathcal{F}).$$

They may be computed as follows. Let $0 \to \mathcal{F} \to \mathcal{J}^\bullet$ be an injective (or flabby) resolution of $\mathcal{F}$. Then

$$H^q_\Phi(X,\mathcal{F}) = H^q(\Gamma_\Phi(X,\mathcal{J}^\cdot)).$$

If Φ is the family cld ($= cld\ (X)$) of closed subsets of X, then we omit the subscript cld. If $A \subset X$ is closed and $\Phi = cld(A)$, then we write $\Gamma_A(X,\mathcal{F})$ and $H^*_A(X,\mathcal{F})$ instead of $\Gamma_{cld(A)}(X,\mathcal{F})$ and $H^*_{cld(A)}(X,\mathcal{F})$. For a subset A of X and a family Φ of supports on X we set $\Phi \cap A := \{Y \cap A \mid Y \subset \Phi\}$ and $\Phi \mid A := \{Y \in \Phi \mid Y \subset A\}$. If A is closed in X, then $\Phi \cap A = \Phi \mid A$.

From now on let X be an *abstract locally semialgebraic space*. We are mainly interested in families of supports on X which are generated by their locally semialgebraic members.

Definition 3. A family Φ of supports on X is called *locally semialgebraic* if every $A \in \Phi$ is contained in some $B \in \Phi \cap \mathcal{T}(X)$.

Now we consider a geometric space M over R. We saw in Chapter I (Prop. 1.4) that the categories of sheaves of Λ-modules on M and $\tilde{M}$ coincide. There is a similar correspondence between the families of supports on M and the locally semialgebraic families of supports on $\tilde{M}$.

Definition 4. Let M be a geometric space over R. A *family of supports* on M is a family $\Phi \subset \bar{\mathcal{T}}(M)$ of closed locally semialgebraic subsets of M such that

a) Every closed locally semialgebraic subset of a member of Φ is in Φ.
b) Φ is closed under finite unions.

A support family Φ on M determines a locally semialgebraic support family $\tilde{\Phi}$ on $\tilde{M}$:

$$\tilde{\Phi} := \{Y \subset \tilde{M} \text{ closed} \mid \text{There is some } A \in \Phi \text{ with } Y \subset \tilde{A}\}.$$

Conversely, if Ψ is a locally semialgebraic family of supports on $\tilde{M}$, then we obtain a family Φ on M as follows:

$$\Phi := \{A \in \bar{T}(M) \mid \tilde{A} \in \Psi\}.$$

Obviously these assignments are inverse to each other. Thus families of supports on M and locally semialgebraic families of supports on $\tilde{M}$ are really the same.
If no confusion seems to be possible, we often do not distinguish between Φ and $\tilde{\Phi}$ in our notation and simply write Φ instead of $\tilde{\Phi}$.

According to our convention in the introduction of this chapter we define

$$\Gamma_\Phi(M, \mathcal{F}) = \Gamma_{\tilde{\Phi}}(\tilde{M}, \tilde{\mathcal{F}}),$$
$$H^*_\Phi(M, \mathcal{F}) = H^*_{\tilde{\Phi}}(\tilde{M}, \tilde{\mathcal{F}}).$$

Of course $H^q_\Phi(M, -), q \geq 0$, are the right derived functors of $\Gamma_\Phi(M, -)$.

Examples 1.1. a) The family $cld(X)$ of closed subsets of the abstract space X is a locally semialgebraic support family.
b) The family $\bar{\gamma}(X)$ of closed semialgebraic subsets of X generates the support family $sa = \{A \subset X \text{ closed} \mid \exists B \in \bar{\gamma}(X) \text{ with } A \subset B\}$. Obviously sa is a locally semialgebraic family.
Now let M be a separated geometric space over R.
c) The family $c(M)$ consisting of all complete semialgebraic subsets of M is a family of supports on M.
d) The support family on M which consists of all partially complete locally semialgebraic subsets is denoted by $pc(M)$.

If there can be no doubt about the space M we simply write c and pc instead of $c(M)$ and $pc(M)$.

Lemma 1.2. Let Φ be a locally semialgebraic family of supports on X and $s \in \Gamma(X, \mathcal{F})$. Then $|s| \in \Phi$ if and only if there exists some $A \in \Phi \cap T(X)$ with $s \mid X \setminus A = 0$.

Proof. Trivial.

We restrict our attention to locally semialgebraic families of supports. This is no serious restriction since most of the interesting families of supports are of this type as the following results indicate.

Lemma 1.3. Let Φ be a family of supports on X. Assume that every $A \in \Phi$ posesses a neighbourhood B in X with $B \in \Phi$ and a closed neighbourhood $C \in \bar{T}(X)$ which is regular and paracompact. Then Φ is locally semialgebraic.

This lemma follows from Remark I.5.6.

Notation: $\Phi^m := \Phi \cap X^{\max}$.

Proposition 1.4. Let Φ be a family of supports on X. We assume that X is maximally closed and $X^{\max}$ is dense in X and that Φ^m is paracompactifying (in the topological sense [B, Def. I.6.1]). Then Φ is locally semialgebraic.

Proof. Let $A \in \Phi$. Then there are sets $B \in \Phi$ and $C \in \Phi$ such that $B^{\max}$ is a neighbourhood of $A^{\max}$ and $C^{\max}$ is a neighbourhood of $B^{\max}$ in $X^{\max}$. Our considerations are now similar to those in the proof of Proposition I.5.7. We choose a locally finite open cover $(U_i \mid i \in I)$ of $C^{\max}$ (U_i open in $C^{\max}$) with the following properties:
i) U_i is contained in some $X_i \in \mathring{\gamma}(X)$.
ii) Either $U_i \subset B^{\max}$ or $U_i \cap A^{\max} = \emptyset$.
iii) If $U_i \cap B^{\max} \neq \emptyset$, then U_i is contained in the interior of $C^{\max}$.

Then we choose open refinements $(V_i \mid i \in I)$ and $(W_i \mid i \in I)$ of $(U_i \mid i \in I)$ with $\bar{V}_i^{\max} \subset W_i \subset \bar{W}_i^{\max} \subset U_i$. Here $\bar{V}_i$ and $\bar{W}_i$ are the closures in X. The sets $\bar{W}_i$ and $\bar{V}_i$ are contained in X_i, hence $\bar{W}_i^{\max}$ and $\bar{V}_i^{\max}$ are compact sets. All coverings being locally finite this means that, for fixed i, W_i meets only finitely many sets $W_j, j \in I$.
Let J be the subset of I consisting of those indices $i \in I$ for which U_i is contained in the interior of $C^{\max}$. Since $\bar{V}_i^{\max}$ is compact we can choose open semialgebraic subsets Y_i of X with $Y_i \subset X_i$ and $V_i \subset Y_i \cap X^{\max} \subset W_i$ for every $i \in J$. By hypothesis $X^{\max}$ is dense in X. Hence $Y_i \cap Y_j$ is empty if $W_i \cap W_j$ is empty. We see that $(Y_i \mid i \in J)$ is a locally finite family in X. Let $J_1 := \{i \in J \mid Y_i \not\subset B\}$. The set

$$D := (\cup(Y_i \mid i \in J)) \setminus (\cup(Y_i \mid i \in J_1))$$

is locally semialgebraic. It is equal to $B \setminus (\cup(Y_i \mid i \in J_1))$, hence it is also closed. By hypothesis $X^{\max}$ is dense in X. Hence $\bar{Y}_i \cap X^{\max} = \overline{Y_i \cap X^{\max}} \cap X^{\max} \subset \bar{W}_i^{\max} \subset U_i$ for every $i \in J$. If $i \in J_1$ then $Y_i \cap X^{\max} \not\subset B^{\max}$ and hence $U_i \cap A^{\max} = \emptyset$ by property ii). We see that $\bar{Y}_i \cap A^{\max}$ and thus also $\bar{Y}_i \cap A$ is empty for $i \in J_1$. Since $(\bar{Y}_i \mid i \in J_1)$ is a locally finite family in X, its union is a closed subset of X which does not meet A. We conclude that D is a closed locally semialgebraic neighbourhood of A in X. Since $D \subset B$ we have $D \in \Phi$, and Prop. 1.4 is proved.

Remark 1.5. The preceding proof even shows that every $A \in \Phi$ has a neighbourhood B in X which belongs to Φ.

Definition 5. A locally semialgebraic family Φ of supports on X is said to be *paracompactifying* if
a) Each element of $\Phi \cap \mathcal{T}(X)$ is regular and paracompact (cf. I, §3).
b) Each element of Φ has a neighbourhood in X which belongs to Φ.

Note the difference to the usual meaning of „paracompactifying" in topology. However, there is the following connection with the classical notion.

Proposition 1.6. Let Φ be a locally semialgebraic family of supports on X.
If Φ is paracompactifying then the induced family Φ^m of supports on $X^{\max}$ is paracompactifying in the classical sense ([B, I, Def. 6.1]). The converse holds if X is maximally closed and $X^{\max}$ is dense in X.

Proof. If Φ is paracompactifying, then $A^{\max}$ is a paracompact topological space for every $A \in \Phi$ (Theorem I.5.7). Moreover, each $A \in \Phi$ has a neighbourhood B in X with $B \in \Phi$. Then $A^{\max}$ has the neighbourhood $B^{\max}$ in $X^{\max}$. We see that Φ^m is paracompactifying in the classical sense.
Conversely, if Φ^m is paracompactifying, X is maximally closed and $X^{\max}$ is dense in X, then Φ satisfies condition b) of Definition 5 as we remarked in 1.5. Every $A \in \Phi \cap \mathcal{T}(X)$ is paracompact and regular by (I.5.7). Hence Φ is paracompactifying.

Examples 1.7. Let M be a separated geometric space over R. The following families of supports are paracompactifying.
i) cld if M is regular and paracompact.
ii) sa if M is regular and taut.
iii) c if M is locally complete.
iv) pc if M is locally complete and paracompact.

Proof. i), ii) are trivial.
iii) If M is locally complete then M is regular ([DK$_3$, I, §7]). Moreover, every complete semialgebraic subset A of M has a complete semialgebraic neighbourhood B in M. It suffices to prove this for M semialgebraic. But then M is affine ([Ro], [DK$_3$, I.4.1]) and our claim follows easily by triangulating M and A simultaneously ([DK$_1$, §2], [DK$_3$, II.4.4]).
iv) M is regular and paracompact. Now use the triangulation theorem ([DK$_3$, II.4.4]) as in the proof of iii).

We make the following *general assumption*:
Unless otherwise stated, up to the end of this book, *all families of supports* on an abstract space X are assumed to be *locally semialgebraic*.

Lemma 1.8. Let Φ be a paracompactifying family of supports on the abstract space X and U be an open subspace of X. Then $\Phi \mid U$ is paracompactifying.

Proof. Let $C \in \Phi$ with $C \subset U$. By assumption C has a neighbourhood E in X with $E \in \Phi$. From Remark I.5.6 we know that there is a neighbourhood $D \in \bar{\mathcal{T}}(X)$ of C in X with $D \subset U \cap E$. We have $D \in \Phi$. This shows that $\Phi \mid U$ is locally semialgebraic and paracompactifiying.

We state another useful fact about families of supports.

Lemma 1.9. Let A be a closed subset of X and $\mathcal{F}$ be a sheaf on A. Assume that Φ is a family of supports on A such that every $B \in \Phi$ has a neighbourhood $C \in \Phi$ (e.g. $\Phi = \Psi \cap A$ with Ψ paracompactifying). Let $s \in \Gamma(A, \mathcal{F})$. Then $\operatorname{supp}(s) \in \Phi$ if and only if $\operatorname{supp}(s \mid A^{\max}) \in \Phi \cap A^{\max}$.

The proof is easy.

§2 - Induced maps and homotopy

We consider pairs $(X,\mathcal{F})$ consisting of an abstract space X and a sheaf $\mathcal{F}$ on X.

Definition 1. A contravariant map between pairs $(X,\mathcal{F})$ and $(Y,\mathcal{G})$ is a pair (f,δ) consisting of a (locally semialgebraic) map $f:X\to Y$ and a homomorphism of sheaves $\delta: f^*\mathcal{G}\to\mathcal{F}$. Up to the end of this section we simply call these maps „maps of pairs".

We may compose two maps $(f,\delta):(X,\mathcal{F})\to(Y,\mathcal{G})$ and $(g,\varepsilon):(Y,\mathcal{G})\to(Z,\mathcal{H})$ as follows: $(g,\varepsilon)\circ(f,\delta):=(g\circ f,\delta\circ f^*(\varepsilon)), g\circ f: X\to Y\to Z, f^*g^*\mathcal{H}\xrightarrow{f^*(\varepsilon)} f^*\mathcal{G}\xrightarrow{\delta}\mathcal{F}$.
In this way we obtain a category of pairs. If $\mathcal{F}=f^*\mathcal{G}$ and $\delta=id$, we simply write f instead of $(f,id):(X,f^*\mathcal{G})\to(Y,\mathcal{G})$. Every map $(f,\delta):(X,\mathcal{F})\to(Y,\mathcal{G})$ induces an homomorphism $(f,\delta)^*: H^*_\Psi(Y,\mathcal{G})\to H^*_\Phi(X,\mathcal{F})$ provided Φ and Ψ are families of supports on X and Y with $f^{-1}(\Psi)\subset\Phi$. Here $f^{-1}(\Psi)$ is the family of supports on X which is generated by $\{f^{-1}(B)\mid B\in\Psi\}$ (i.e. it consists of those closed subsets of X which are contained in some $f^{-1}(B), B\in\Psi$). The induced homomorphism may be described as follows, cf. [B, II.8]:
Let $0\to\mathcal{G}\to\mathcal{J}_1^\bullet$ and $0\to\mathcal{F}\to\mathcal{J}_2^\bullet$ be injective resolutions of $\mathcal{G}$ and $\mathcal{F}$ on Y and X respectively. Then $0\to f^*\mathcal{G}\to f^*\mathcal{J}_1^\bullet$ is a resolution of $f^*\mathcal{G}$ and there is a morphism α of resolutions, unique up to chain homotopy, such that the diagram

$$\begin{array}{ccc} f^*\mathcal{J}_1^\bullet & \xrightarrow{\alpha} & \mathcal{J}_2^\bullet \\ \uparrow & & \uparrow \\ f^*\mathcal{G} & \xrightarrow[\delta]{} & \mathcal{F} \end{array}$$

commutes. Now $(f,\delta)^*$ is induced by the chain map

$$\Gamma_\Psi(Y,\mathcal{J}_1^\bullet)\to\Gamma_{f^{-1}(\Psi)}(X,f^*\mathcal{J}_1^\bullet)\xrightarrow{\alpha}\Gamma_\Phi(X,\mathcal{J}_2^\bullet).$$

The semialgebraic subspace $\{x\in \operatorname{Sper}\mathbb{Z}[T]\mid 0\le T\le 1\}$ of $\operatorname{Sper}\mathbb{Z}[T]$ is called the *unit interval* and is denoted by I. For every abstract space X we denote the product of X and I in the category of abstract spaces by $X\times I$. (The product is taken over the final object $\operatorname{Sper} R_0$ in the category of abstract spaces, R_0 the field of real algebraic numbers). This product exists ([S$_1$, II.3]). If X is a semialgebraic subspace of a real spectrum $\operatorname{Sper} A$ it may be described as follows:
Let $p:\operatorname{Sper} A[T]\to\operatorname{Sper} A$ be the projection (which is induced by the natural embedding $A\to A[T]$). Then $X\times I$ is the semialgebraic subspace $\{x\in\operatorname{Sper} A[T]\mid p(x)\in X, 0\le T(x)\le 1\}$ of $\operatorname{Sper} A[T]$. The projection $X\times I\to X$ is denoted by π. In the special case we just described π is the restriction of p to $X\times I\subset\operatorname{Sper} A[T]$. The constructible subsets $\{T=0\}$ and $\{T=1\}$ of $\operatorname{Sper}\mathbb{Z}[T]$ are one point sets $\{P_0\}$ and $\{P_1\}$. The constant maps $p_0:X\to I$ resp. $p_1:X\to I$ mapping X to $\{P_0\}$ resp. $\{P_1\}$ are locally semialgebraic maps. Thus we obtain the locally semialgebraic maps

$$i_0:=(id,p_0), i_1:=(id,p_1): X\rightrightarrows X\times I.$$

i_0 and i_1 are the embeddings of X onto the bottom and the top of $X\times I$.
If X is a subspace of $\operatorname{Sper} A$, then i_0 (resp. i_1) is the restriction to X of the map $\operatorname{Sper} A\to\operatorname{Sper} A[T]$ which is induced by the ring homomorphism $A[T]\to A, T\mapsto 0$ (resp. $T\mapsto 1$).

Example 2.1. Let M be a geometric space over R. Since the functor $\sim$ commutes with fibre products ([S$_1$, III.2.1]) we conclude that

$$\tilde{M} \times I = (M \times [0,1])^{\sim}$$

where [0,1] is the unit interval in R.

Let Φ be a family of supports on X and $\mathcal{F}$ be a sheaf on X. Then $\Phi \times I$ denotes the support family $\pi^{-1}(\Phi)$ on $X \times I$ and $\mathcal{F} \times I$ the sheaf $\pi^*\mathcal{F}$ on $X \times I$.

Definition 2 (cf. [D$_2$, §4]). Let $(f,\delta),(g,\varepsilon):(X,\mathcal{F}) \rightrightarrows (Y,\mathcal{G})$ be maps between pairs. Let Φ and Ψ be families of supports on X and Y. Then (f,δ) and (g,ε) are said to be *homotopic with respect to* Φ *and* Ψ (written $(f,\delta) \underset{\Phi,\Psi}{\simeq} (g,\varepsilon)$) if there is a map $(H,\Theta):(X\times I,\mathcal{F}\times I) \to (Y,\mathcal{G})$ (a „homotopy") such that $H^{-1}(\Psi) \subset \Phi \times I$ and $(H,\Theta) \circ i_0 = (f,\delta), (H,\Theta) \circ i_1 = (g,\varepsilon)$. If $\Phi = cld$ and $\Psi = cld$ then (f,δ) and (g,ε) are simply called homotopic (written $(f,\delta) \simeq (g,\varepsilon)$).

Example 2.2. Let $f,g:M \to N$ be locally semialgebraic maps between geometric spaces over R. Then we may consider f and g as maps of pairs $(M,G_M) \to (N,G_N)$ for any Λ-module G.
Since $M \times [0,1] = \tilde{M} \times I$ (Ex. 2.1) we see that $f \simeq g$ if there is a locally semialgebraic map $H:M \times [0,1] \to N$ with $H(-,0) = f$ and $H(-,1) = g$. We have $f \underset{sa}{\simeq} g \quad (f \underset{c}{\simeq} g, f \underset{pc}{\simeq} g)$ if the homotopy H is a semialgebraic (proper semialgebraic, partially proper) map (cf. [DK$_3$, I, §5, §6]). Here we used the abbreviations $\underset{sa}{\simeq} := \underset{sa,sa}{\simeq}$, $\underset{c}{\simeq} := \underset{c,c}{\simeq}$, $\underset{pc}{\simeq} := \underset{pc,pc}{\simeq}$.

We recall two results from the paper [D$_2$]. Let $\mathcal{F}$ be a sheaf on the abstract space X.

Proposition 2.3. ([D$_2$, Prop. 4.4]). The adjunction homomorphism

$$\mathcal{F} \longrightarrow \pi_*\pi^*\mathcal{F} = \pi_*(\mathcal{F} \times I)$$

is an isomorphism.

Proposition 2.4 ([D$_2$, Prop. 4.2]). The higher direct images $R^q\pi_*(\mathcal{F} \times I)$ are zero for $q > 0$.

Now it is easy to derive

Theorem 2.5 (cf. [D$_2$, Thm. 4.1]). Let $(f,\delta),(g,\varepsilon):(X,\mathcal{F}) \rightrightarrows (Y,\mathcal{G})$ be maps between pairs. Let Φ and Ψ be families of supports on X and Y. Assume that $(f,\delta) \underset{\Phi,\psi}{\simeq} (g,\varepsilon)$. Then the induced maps in cohomology coincide:

$$(f,\delta)^* = (g,\varepsilon)^* : H^*_\Psi(Y,\mathcal{G}) \longrightarrow H^*_\Phi(X,\mathcal{F}).$$

Proof. Let (H,Θ) be a homotopy from (f,δ) to (g,ε). Since $(f,\delta)^* = i_0^* \circ (H,\Theta)^*$ and $(g,\varepsilon)^* = i_1^* \circ (H,\Theta)^*$ it suffices to prove the assertion for the maps $i_0, i_1:(X,\mathcal{F}) \rightrightarrows (X \times I, \mathcal{F} \times I)$ which are homotopic with respect to Φ and $\Phi \times I$. We consider the Leray spectral sequence $E_2^{pq} = H^p_\Phi(X, R^q\pi_*(\mathcal{F}\times I)) \Longrightarrow H^{p+q}_{\Phi\times I}(X\times I, \mathcal{F}\times I) = E^{p+q}$ ([B, IV.6]). From Prop. 2.4 we conclude that this sequence splits. Hence the edge homomorphism

$$\lambda\colon H^p_\Phi(X, \pi_*(\mathcal{F}\times I)) \longrightarrow H^p_{\Phi\times I}(X\times I, \mathcal{F}\times I)$$

is an isomorphism for every $p\geq 0$.

Prop. 2.3 implies that

$$H^p_\Phi(X, \mathcal{F}) \xrightarrow{\ \eta\ } H^p_\Phi(X, \pi_*(\mathcal{F}\times I))$$

is an isomorphism for every $p\geq 0$. Hence $\lambda\circ\eta\colon H^p_\Phi(X,\mathcal{F}) \longrightarrow H^p_{\Phi\times I}(X\times I, \mathcal{F}\times I)$ is an isomorphism. But $\lambda\circ\eta$ is just the map π^* which is induced by $\pi\colon (X\times I, \mathcal{F}\times I)\longrightarrow (X,\mathcal{F})$ ([B, IV.6.3]). Thus π^* is an isomorphism. Since $\pi\circ i_0 = \pi\circ i_1 = id_X$ we finally conclude that

$$i_0^* = (\pi^*)^{-1} = i_1^*.$$

q.e.d.

In the following we give a useful application of the „homotopy axiom" 2.5. We show that important families of supports on a geometric space can be substituted by paracompactifying ones provided the coefficient sheaves are locally constant.

So let us consider a geometric space M over R and two closed locally semialgebraic subsets $A_1 \subset A_2$ of M. Furthermore, we assume that there is some $C\in \mathring{\mathcal{T}}(M)$ such that C is a strong deformation retract of both sets $M\setminus A_1$ and $M\setminus A_2$ (cf. [DK$_3$, III, §1]). Let

$$F_k\colon (M\setminus A_k)\times[0,1] \longrightarrow M\setminus A_k$$

be a strong deformation retraction (loc. cit.) with $F_k(-,0) = id$, $F_k(-,1) =: r_k$ a retraction from $M\setminus A_k$ to C and $F_k(x,t) = x$ for every $x\in C$ and every $t\in[0,1]$ $\quad(k=1,2)$. Let $\mathcal{F}$ be a sheaf on M such that $F_k^*(\mathcal{F}\mid M\setminus A_k) = (\mathcal{F}\mid M\setminus A_k)\times I$ for $k=1,2$ (e.g. a constant sheaf).

Lemma 2.6. In this situation the canonical map

$$H^*_{A_1}(M,\mathcal{F}) \longrightarrow H^*_{A_2}(M,\mathcal{F})$$

is an isomorphism.

Proof. Let $i_k\colon C\hookrightarrow M\setminus A_k$ be the inclusion map $(k=1,2)$. From Theorem 2.5 we derive that $H^*(M\setminus A_k,\mathcal{F}) \xrightarrow{i_k^*} H^*(C,\mathcal{F}) \xrightarrow{r_k^*} H^*(M\setminus A_k,\mathcal{F})$ is the identity map. On the other hand $r_k\circ i_k = id_C$, hence $i_k^*\circ r_k^*$ is also the identity map. We see that both maps

$$H^*(M\setminus A_k,\mathcal{F}) \longrightarrow H^*(C,\mathcal{F}) \quad (k=1,2)$$

are isomorphisms. Hence

$$H^*(M\setminus A_1,\mathcal{F}) \longrightarrow H^*(M\setminus A_2,\mathcal{F})$$

is also an isomorphism.

Now we consider the long exact cohomology sequences (S_k)

$$\dots \to H^n_{A_k}(M,\mathcal{F}) \to H^n(M,\mathcal{F}) \to H^n(M\setminus A_k,\mathcal{F}) \to H^{n+1}_{A_k}(M,\mathcal{F}) \to \dots$$

for $k = 1,2$ (cf. [B, II.12]). There is a canonical homomorphism $(S_1) \to (S_2)$ and we can compare both sequences. The Five Lemma yields the desired result.

Example 2.7. Let M be a locally finite simplical complex over R (cf. [DK$_3$, I, §2]) and let $A \subset M$ be a closed subcomplex. We denote the closure $\overline{St_{M''}(A)} \cap M$ of the open star neighbourhood $St_{M''}(A)$ of A in the second barycentric subdivision M'' of M by A_2. Let $C := co_{M'}(M\setminus A)$ be the core of $M\setminus A$ in the first barycentric subdivision M' of M ([DK$_3$, III, §1]). C is the largest partially complete subcomplex of M' which is contained in $M\setminus A$. Obviously $C \subset M\setminus A_2$ and the star neighbourhood $V := St_{M'}(C)$ of C in M' is just $M\setminus A$. We have a canonical retraction

$$r : V = M\setminus A \longrightarrow C$$

with the following property:

If $x \in V$ is contained in an open simplex S of M', then the whole open line segment $]x, r(x)[$ is contained in S (loc. cit.).

We recall the explicit definition of r. Let $S \subset V$ be an open simplex of M' spanned by the vertices $e_0,\dots,e_n$. Assume that $e_l,\dots,e_n$ are those vertices of S which are contained in $M\setminus A$. Then $\bar{S}\cap C$ is the closed face $[e_l,\dots,e_n]$ of $\bar{S}$ spanned by $e_l,\dots,e_n$. If $x = \sum_{i=0}^{n} \lambda_i e_i \in S$ $\quad(\lambda_i \in R, \lambda_i > 0, \sum_{i=0}^{n}\lambda_i = 1)$ then

$$r(x) := \Big(\sum_{i=l}^{n} \lambda_i e_i\Big) \Big/ \Big(\sum_{i=l}^{n}\lambda_i\Big).$$

Lemma 2.8. If $x \in M\setminus A_2$, then $]x, r(x)[$ is contained in $M\setminus A_2$.

Proof. As above let $S \subset V$ be an open simplex of M' and $x \in S\setminus A_2$. We use the notation from above. Consider $y := (1-t)\cdot x + t\cdot r(x)$ $\quad (t \in]0,1[)$. Let $T :=]e_{i_0},\dots,e_{i_m}[$ be any open face of S which is contained in A. Then all indices i_k are $< l$. Since $x \notin A_2$, there is some $\lambda_k, k \notin \{i_0,\dots,i_m\}$, with $\lambda_k > \min\{\lambda_{i_0},\dots,\lambda_{i_m}\}$ ([DK$_2$, 2.6]). Now $y = \sum_{i=0}^{n} \mu_i e_i$ with

$$\mu_i = \begin{cases} (1-t)\cdot\lambda_i & \text{if } \quad 0 \le i < l \\ (1-t)\cdot\lambda_i + \rho_i, \rho_i > 0, & \text{if } \quad l \le i \le n. \end{cases}$$

We see that $\mu_k > \min\{\mu_{i_0},\dots,\mu_{i_m}\}$. Applying [DK$_2$, 2.6] again we conclude that $y \notin \overline{St_{M''}(T)}$. Since $A_2 \cap S$ is the union of the sets $\overline{St_{M''}(T)} \cap S$ with T running through the faces $T \subset A$ of S, the point y is not contained in A_2. q.e.d.

Now we define a deformation retraction $F_1 : (M\setminus A) \times [0,1] \longrightarrow M\setminus A$ by $F_1(x,t) := (1-t)\cdot x + t\cdot r(x)$. According to Lemma 2.8 F_1 restricts to a deformation retraction

$$F_2 : (M\setminus A_2) \times [0,1] \longrightarrow M\setminus A_2$$

Let $\mathcal{F}$ be a locally constant sheaf on M. Then $\mathcal{F}$ is constant on all star-shaped subsets of M. This implies that

$$F_k^*(\mathcal{F} \mid M \setminus A_k) = (\mathcal{F} \mid M \setminus A_k) \times I \quad (k = 1, 2; A_1 := A).$$

Thus we finally obtain from Lemma 2.6

$$H_A^*(M, \mathcal{F}) \xrightarrow{\sim} H_{A_2}^*(M, \mathcal{F}).$$

Proposition 2.9. Let M be a paracompact and regular space over R and $\mathfrak{A}$ be a finite family in $\bar{\mathcal{T}}(M)$ which is closed under intersections (i.e. $A \cap B \in \mathfrak{A}$ if $A, B \in \mathfrak{A}$). Then there are paracompactifying families $\Phi(A)$ of supports ($A \in \mathfrak{A}$) such that $A \in \Phi(A), \Phi(A) \cap \Phi(B) = \Phi(A \cap B)$ for $A, B \in \mathfrak{A}$ and the canonical map

$$H_A^*(M, \mathcal{F}) \longrightarrow H_{\Phi(A)}^*(M, \mathcal{F})$$

is an isomorphism for every locally constant sheaf $\mathcal{F}$ on M and every $A \in \mathfrak{A}$. If closed locally semialgebraic neighbourhoods $W(A)$ of A ($A \in \mathfrak{A}$) are given, then we may choose $\Phi(A)$ as a subfamily of $\bar{\mathcal{T}}(W(A))$.

Proof. We may assume that M is a locally finite simplical complex and that A and $W(A)$ are subcomplexes for every $A \in \mathfrak{A}$ ([DK$_3$, II, 4.4]). Let M^k be the k-th barycentric subdivision of M. Consider some $A \in \mathfrak{A}$. We set $A_1 := A$ and

$$A_{n+1} := \overline{St_{M^{2n}}(A_n)} \cap M \quad (n = 1, 2, \ldots).$$

Let $\Phi(A)$ be the union of the families $\bar{\mathcal{T}}(A_n), n \in \mathbb{N}$. Then $\Phi(A)$ is a paracompactifying family of supports and it is contained in $\bar{\mathcal{T}}(W(A))$. Since $H_{A_n}^*(M, \mathcal{F}) \xrightarrow{\sim} H_{A_{n+1}}^*(M, \mathcal{F})$ by Example 2.7, we have $H_A^*(M, \mathcal{F}) \xrightarrow{\sim} H_{\Phi(A)}^*(M, \mathcal{F})$.
By induction we conlude from the following Lemma 2.10 that $\Phi(A) \cap \Phi(B) = \Phi(A \cap B)$.

Lemma 2.10. $\overline{St_{M^2}(A)} \cap \overline{St_{M^2}(B)} \cap M = \overline{St_{M^2}(A \cap B)} \cap M$.

Proof. Let σ be a simplex of M^2 with $\sigma \subset \overline{St_{M^2}(A)} \cap \overline{St_{M^2}(B)}$. Let S be the open simplex of M^1 containing σ. Since either $\bar{S} \cap A \subset \bar{S} \cap B$ or $\bar{S} \cap B \subset \bar{S} \cap A$, we assume without loss of generality that $\bar{S} \cap A \subset \bar{S} \cap B$. Now we choose some $\tau \subset St_{M^2}(A)$ with $\sigma \leq \tau$ ($\leq$ means „σ is a face of τ"). τ is spanned by the barycentres $\hat{T}_0, \ldots, \hat{T}_n$ of some simplices $T_0 < \ldots < T_n$ of M^1. Let $\hat{T}_{i_0}, \ldots, \hat{T}_{i_m}$ be the vertices of σ. Necessarily $S = T_{i_m}$. There is some T_i with $T_i \subset A$. We choose i minimal with $T_i \subset A$. Since A is closed in M and $T_{i_m} = S$ ist contained in M, we have $i \leq i_m$. Hence $T_i \subset \bar{S} \cap A \subset \bar{S} \cap B$ and we see that $\tau \in St_{M^2}(A \cap B)$. This proves Lemma 2.10.

§3 - Extension of sections to a neighbourhood

We show that in many cases sections of a sheaf over a subspace can be extended into a neighbourhood.
Let X be an abstract space.

Lemma 3.1. Let $A \subset B$ be subsets of X. Assume that A is partially quasicompact (cf. Def. 1, I, §5) and posesses a paracompact and regular locally semialgebraic neighbourhood in X. Let $\mathcal{F}$ be a sheaf on B. Then the canonical map

$$\varinjlim_{A \subset U \in \mathring{T}(X)} \Gamma(U \cap B, \mathcal{F}) \xrightarrow{\alpha} \Gamma(A, \mathcal{F})$$

is an isomorphism.

Proof. The family $(U \cap B \mid U \in \mathring{T}(X), A \subset U)$ is a fundamental system of neighbourhoods of A in B (I.5.5), hence α is certainly injective.

To prove that α is surjective we may clearly assume that X is paracompact and regular. Let $(X_i \mid i \in I)$ be a locally finite open covering of X by sets $X_i \in \mathring{\gamma}(X)$. We choose refinements $(W_i \mid i \in I)$ and $(B_i \mid i \in I)$ of $(X_i \mid i \in I)$ with $W_i \in \mathring{\gamma}(X), B_i \in \bar{\gamma}(X)$ and $W_i \subset B_i \subset X_i$ for every $i \in I$ as described in Prop. I.5.1. Consider a section $s \in \Gamma(A, \mathcal{F})$ of $\mathcal{F}$ over A. There are coverings $(U_j \mid j \in J(i))$ of $A \cap B_i$ by open semialgebraic subsets U_j of X_i and sections $s_j \in \Gamma(U_j \cap B, \mathcal{F}), j \in J(i)$, such that $s_j \mid U_j \cap A = s \mid U_j \cap A$. Since $A \cap B_i$ is quasicompact, we may assume that $J(i)$ is finite for every $i \in I$. Let $J := \cup(J(i) \mid i \in I)$ be the disjoint union of the index sets $J(i)$. Then $(U_j \mid j \in J)$ is a locally finite family, hence $U := \cup(U_j \mid j \in J)$ is an element of $\mathring{T}(X)$. Now we shrink the covering $(U_j \mid j \in J)$ of U, i.e. we choose a covering $(C_j \mid j \in J)$ of U by closed semialgebraic subsets $C_j \in \bar{\gamma}(U)$ of U such that $C_j \subset U_j$ for every $j \in J$.
Let

$$W := \{x \in U \cap B \mid \quad \text{If } x \in C_i \cap C_j, \quad \text{then } (s_i)_x = (s_j)_x\}.$$

It is easy to check that W is an open neighbourhood of A in B. Using Lemma I.5.5 again we take some $V \in \mathring{T}(X)$ with $A \subset V \cap B \subset W$ and define $t \in \Gamma(V \cap B, \mathcal{F})$ by $t \mid V \cap B \cap C_j = s_j \mid V \cap B \cap C_j$. Then $\alpha(t) = s$. q.e.d.

Proposition 3.2. Let $A \subset B$ be subsets of X and Φ be a paracompactifying family of supports on X. Assume that $C \cap A$ is partially quasicompact for every $C \in \Phi$. Let $\mathcal{F}$ be a sheaf on B. Then the canonical map

$$\varinjlim_{A \subset U \in \mathring{T}(X)} \Gamma_{\Phi \cap B \cap U}(U \cap B, \mathcal{F}) \xrightarrow{\alpha} \Gamma_{\Phi \cap A}(A, \mathcal{F})$$

is an isomorphism.

Proof. We first prove injectivity. Let $s \in \Gamma_{\Phi \cap B \cap U}(U \cap B, \mathcal{F}), A \subset U \in \mathring{T}(X)$, be a section with $s \mid A = 0$. Let C be an element of $\Phi \cap T(X)$ such that $\operatorname{supp}(s)$ is contained in $C \cap U \cap B$. According to the preceding Lemma 3.1 there is some $V \in \mathring{T}(X), V \subset U$, such that $C \cap A \subset V$ and $s \mid V \cap B = 0$. Then $W := V \cup (U \setminus C) \in \mathring{T}(X)$ and $s \mid W \cap B = 0$.

Hence α is injective.
Now we consider a section $s \in \Gamma_{\Phi\cap A}(A,\mathcal{F})$. Let C, D be elements of $\Phi\cap\mathcal{T}(X)$ such that D is a neighbourhood of C in X and $\operatorname{supp}(s) \subset C\cap A$. By Lemma 3.1 we find some neighbourhood $V \in \mathring{\mathcal{T}}(X)$ of $D\cap A$ and a section $t_1 \in \Gamma(V\cap B,\mathcal{F})$ with $t_1 \mid D\cap A = s \mid D\cap A$. Let ∂D be the boundary of D in X. Since $t_1 \mid \partial D\cap A = 0$ there is a neighbourhood $W \in \mathring{\mathcal{T}}(X)$ of $\partial D \cap A$ such that $W \subset V$ and $t_1 \mid W\cap B = 0$ (Lemma 3.1). Now we choose some neighbourhood $U_1 \in \mathring{\mathcal{T}}(X)$ of $(D\setminus W)\cap A$ with $U_1 \subset (D\setminus\partial D)\cap V$. This is possible by Lemma I.5.5. Let $U_2 = U_1\cup W$ and $t_2 = t_1 \mid U_2\cap B$. Since $t_2 \mid (U_2\setminus D)\cap B = 0$ we may extend t_2 by zero to a section $t \in \Gamma(U\cap B,\mathcal{F})$ where $U = U_2\cup(X\setminus D)$. Clearly $A\subset U, \operatorname{supp}(t) \subset D\cap B\cap U \in \Phi\cap B\cap U$ and $\alpha(t) = t \mid A = s$. Thus α is also surjective.

Let Φ be a paracompactifying family of supports on X.

Notation. We denote by X_Φ the union

$$\cup(A\in\Phi \mid A\in\mathcal{T}(X))$$

of all locally semialgebraic elements of Φ. X_Φ is a locally semialgebraic subspace of X (cf. Chapter I, §2). For a subset A of X and a sheaf $\mathcal{F}$ on A we denote by $\mathcal{F}_\Phi$ the restriction of $\mathcal{F}$ to $X_\Phi\cap A$.

The following two statements are obvious.

Lemma 3.3. Let Φ be a paracompactifying family of supports on X, A be a subset of X and $\mathcal{F}$ be a sheaf on A. Then

$$\Gamma_{\Phi\cap A}(A,\mathcal{F}) = \Gamma_{\Phi\cap A}(A\cap X_\Phi,\mathcal{F}_\Phi)$$
$$H^*_{\Phi\cap A}(A,\mathcal{F}) = H^*_{\Phi\cap A}(A\cap X_\Phi,\mathcal{F}_\Phi).$$

Lemma 3.4. Let Φ be a paracompactifying family of supports on X. Then X_Φ is a taut and regular space.

Corollary 3.5. Let Φ be a paracompactifying family of supports on $X, A\subset X$ a closed subset and $\mathcal{F}$ a sheaf on A. Then

$$\Gamma_{\Phi\cap A}(A,\mathcal{F}) = \Gamma_{\Phi\cap A^{\max}}(A^{\max},\mathcal{F}).$$

Proof. Replacing X by X_Φ we may assume that X is taut and hence maximally closed. Then Corollary 3.5 is an immediate consequence of Proposition 3.2 since A is the only open neighbourhood of $A^{\max}$ in A.

Example 3.6. Corollary 3.5 does not hold for non paracompactifying families of supports. Consider e.g. the affine semialgebraic space $M := [-1,1]$ over R and the sheaf $\mathcal{F}$ on M which is associated to the presheaf

$$(M\cap]a,b[) \longrightarrow \begin{cases} 0 & \text{if} \quad a>0, b>0 \\ \mathbb{Z} & \text{else} \end{cases}$$

Let $A := [-1,0] \in \bar{\gamma}(M)$ and $\Phi := \bar{\gamma}(A)$. Then $\Gamma_\Phi(M,\mathcal{F}) = \Gamma_{\tilde{A}}(\tilde{M},\mathcal{F}) = 0$ and $\Gamma_{\Phi^m}(\tilde{M}^{\max},\mathcal{F}) = \Gamma_A([-1,1]_{top},\mathcal{F}) = \mathbb{Z}$.

Remark 3.7. Let X be a regular and taut space and A be a closed subset of X. The canonical retraction $r_X : X \longrightarrow X^{\max}$ (I.3.12 a) is continuous (I.4.3) and yields, by restriction, a continuous retraction $r_A : A \longrightarrow A^{\max}$. Carral and Coste observed that $\mathcal{F} \mid A^{\max}$ coincides with the direct image $(r_A)_*\mathcal{F}$ of $\mathcal{F}$ for every sheaf $\mathcal{F}$ on A ([CC, Prop. 4]). This is easy to see: If W is open in $A^{\max}$ then $r_A^{-1}(W)$ is open in A and it is contained in every open subset U of A which contains W. Hence $\varinjlim_{W \subset U} \Gamma(U, \mathcal{F}) = \Gamma(r_A^{-1}(W), \mathcal{F}) = \Gamma(W, (r_A)_*\mathcal{F})$ and we see that $\mathcal{F} \mid A = (r_A)_*\mathcal{F}$. This fact means in particular that

$$\Gamma(A, \mathcal{F}) = \Gamma(A^{\max}, \mathcal{F}).$$

Now let Ψ be a family of supports on A such that every $B \in \Psi$ posesses a neighbourhood $C \in \Psi$. Then $\operatorname{supp}(s) \in \Psi$ if and only if $\operatorname{supp}(s \mid A^{\max}) \in \Psi \cap A^{\max}$ (Lemma 1.9). Thus

$$\Gamma_\Psi(A, \mathcal{F}) = \Gamma_{\Psi \cap A^{\max}}(A^{\max}, \mathcal{F}),$$

and we have obtained a different proof (of a slightly refined version) of Corollary 3.5.

§4 - Acyclic sheaves

Let X be a topological space and Φ be a family of supports on X. A sheaf $\mathcal{F}$ on X is called Φ-acyclic if $H^q_\Phi(X,\mathcal{F}) = 0$ for every $q > 0$ ([B, II.4]). The cohomology groups $H^q_\Phi(X,\mathcal{G})$ of X with coefficients in a sheaf $\mathcal{G}$ can be computed by means of any Φ-acyclic resolution of $\mathcal{G}$ ([B, II.4.1], [G, II.4.7.1]). Recall the following definitions from classical sheaf theory ([B, II.5.9], [G, II.3.1, 3.5]).

Definition 1. a) A sheaf $\mathcal{F}$ on X is *flabby* if the restriction map $\Gamma(X,\mathcal{F}) \longrightarrow \Gamma(U,\mathcal{F})$ is surjective for every open subset U of X.
b) A sheaf $\mathcal{F}$ on X is called Φ-*soft* if the restriction map $\Gamma(X,\mathcal{F}) \longrightarrow \Gamma(A,\mathcal{F})$ is surjective for every $A \in \Phi$. If Φ consists of all closed subsets of X, then $\mathcal{F}$ is simply called *soft*.

Flabby sheaves are Φ-acyclic for any family Φ of supports. Φ-soft sheaves are Φ-acyclic if Φ is paracompactifying in the classical sense ([B, I, Def. 6.1], [B, II.5.9], [G, II.4.7.1]). We will prove analogous results in our (locally semialgebraic) setting.

Proposition 4.1. Assume that every $A \in \Phi$ has a neighbourhood B in X with $B \in \Phi$. Then the following statements are equivalent for a sheaf $\mathcal{F}$ on X.

i) $\mathcal{F}$ is Φ-soft.
ii) $\mathcal{F} \mid A$ is soft for every $A \in \Phi$.
iii) $\Gamma_\Phi(X,\mathcal{F}) \longrightarrow \Gamma_{\Phi|A}(A,\mathcal{F})$ is surjective for every closed subset A of X.

If X is an abstract locally semialgebraic space and Φ is locally semialgebraic, then i) - iii) are also equivalent to

iv) $\mathcal{F} \mid A$ is soft for every $A \in \Phi \cap \mathcal{T}(X)$.

Proof. The equivalence of i), ii) and iii) is shown in [B, II, 9.3]. (Our hypothesis is sufficient in the proof given there). If X is an abstract space and Φ is locally semialgebraic, then the equivalence of ii) and iv) is obvious since every $A \in \Phi$ is contained in some $B \in \Phi \cap \mathcal{T}(X)$.

In the following X is an *abstract locally semialgebraic space* and Φ is a family of supports on X. (As always , Φ is assumed to be locally semialgebraic, cf. §1).

Proposition 4.2. Suppose that each $A \in \Phi \cap \mathcal{T}(X)$ is regular and paracompact. Then a sheaf $\mathcal{F}$ on X is Φ-soft if and only if $\Gamma(X,\mathcal{F}) \longrightarrow (A,\mathcal{F})$ is surjective for every $A \in \Phi\cap\mathcal{T}(X)$.

Proof. We only have to prove the „if-part". Let $B \in \Phi$ and $s \in \Gamma(B,\mathcal{F})$. Since Φ is locally semialgebraic, there is some set $C \in \Phi \cap \mathcal{T}(X)$ containing B. By hypothesis C is regular and paracompact. Lemma 3.1 says that s can be extended to a section $s_1 \in \Gamma(U,\mathcal{F})$ of $\mathcal{F}$ over a neighbourhood U of B *in* C. By Remark I.5.6 there is a closed locally semialgebraic subset $A \in \bar{\mathcal{T}}(C)$ of C with $B \subset A \subset U$. Since $C \in \Phi$ we have $A \in \Phi$. Thus the restriction map $\Gamma(X,\mathcal{F}) \longrightarrow (A,\mathcal{F})$ is surjective. We see that $s_1 \mid A$ and hence s can be extended to a global section $t \in \Gamma(X,\mathcal{F})$.

An important class of sheaves consists of the sheaves $\mathcal{F}$ which are flabby with respect to open locally semialgebraic subsets.

Definition 2. Let Y be a subset of X. A sheaf $\mathcal{F}$ on Y is called sa-flabby („semialgebraically flabby") if the restriction map $\Gamma(Y,\mathcal{F}) \longrightarrow (U \cap Y,\mathcal{F})$ is surjective for every $U \in \mathring{\mathcal{T}}(X)$.

Remark 4.3. A flabby sheaf $\mathcal{F}$ is clearly *sa*-flabby. The converse is false in general. Consider for example the semialgebraic space $M := \mathbf{R}$ over $\mathbf{R}$ and the sheaf $\mathcal{F}$ on M whose sections over $U \in \mathring{\gamma}(M)$ are the (set-theoretical) maps $f : U \longrightarrow \mathbf{Z}$ with $f(x) = 0$ for all but finitely many $x \in U$. Then $\tilde{\mathcal{F}}$ is clearly *sa*-flabby, but $\tilde{\mathcal{F}}$ is not flabby. Take e.g. the open subset $V := \widetilde{\bigcup(]2n-1, 2n[\mid n \in \mathbb{N})}$ of $\tilde{M}$. Then $\Gamma(V, \tilde{\mathcal{F}}) = \prod_{n \in \mathbb{N}} \Gamma(]2n-1, 2n[, \mathcal{F})$. Obviously the restriction map $\Gamma(\tilde{M}, \tilde{\mathcal{F}}) = \Gamma(M, \mathcal{F}) \longrightarrow \Gamma(V, \tilde{\mathcal{F}})$ is not surjective.

Following our convention in the introduction of this chapter a sheaf $\mathcal{F}$ on a geometric space M over R is said to be flabby (*sa*-flabby, Ψ-soft) if and only if the sheaf $\tilde{\mathcal{F}}$ on $\tilde{M}$ is flabby (*sa*-flabby, Ψ-soft). (Here Ψ is a support family on M).

Lemma 4.4. Suppose X is regular and taut. Let A be a closed subset of X and $\mathcal{G}$ be a sheaf on A. Then $\mathcal{G}$ is $(\Phi \cap A)$-soft if and only if the restriction $\mathcal{G} \mid A^{\max}$ is $(\Phi \cap A^{\max})$-soft.

Proof. We know from Remark 3.7 that $\Gamma(B, \mathcal{G}) = \Gamma(B^{\max}, \mathcal{G})$ for every closed subset B of A. Lemma 4.4 follows.

„Soft" is a local property. Let $\mathcal{F}$ be a sheaf on X.

Lemma 4.5. We assume that X is regular and paracompact and that every $x \in X$ posesses a closed neighbourhood A such that $\mathcal{F} \mid A$ is soft. Then $\mathcal{F}$ is soft.

Proof. $X^{\max}$ is a paracompact topological space (I.5.7). Now the result follows from [B, II.15.5] (or [G, II.3.4.1]) and Lemma 4.4.

Proposition 4.6. Let Φ be paracompactifying. Assume that every $x \in \bigcup(A \mid A \in \Phi)$ has a closed neighbourhood B such that $\mathcal{F} \mid B$ is soft. Then $\mathcal{F}$ is soft.
{ N.B. The hypothesis of Prop. 4.6 is satisfied if there is an open covering $(X_i \mid i \in I)$ of X such that $\mathcal{F} \mid X_i$ is $(\Phi \mid X_i)$-soft. (Use Proposition I.5.1) }.

Proof. By 4.1 iv) we have to prove that $\mathcal{F} \mid A$ is soft for every $A \in \Phi \cap \mathcal{T}(X)$. Now Lemma 4.5 gives the desired result.

Under certain restrictions „*sa*-flabby" is also a local property.

Definition 3. A space X is said to be of type (L) if X is the disjoint union of a family $(X_i \mid i \in I)$ of open locally semialgebraic subsets $X_i \in \mathring{\mathcal{T}}(X)$ such that every X_i is Lindelöf.

Note that X is of type (L) in the following cases:

a) X is Lindelöf.
b) X is paracompact and connected (I.3.6).
c) $X = \tilde{M}$ and M is a geometric space over R all whose connected components are Lindelöf (e.g. M paracompact).

Proposition 4.7. Suppose X is of type (L). Let $\mathcal{F}$ be a sheaf on X and $(U_j \mid j \in J)$ be an open cover of X by locally semialgebraic subsets $U_j \in \mathring{\mathcal{T}}(X)$ such that $\mathcal{F} \mid U_j$ is *sa*-flabby for every $j \in J$. Then $\mathcal{F}$ is *sa*-flabby.

In particular, a sheaf $\mathcal{G}$ on X is sa-flabby if and only if $\Gamma(X,\mathcal{G}) \longrightarrow \Gamma(U,\mathcal{G})$ is surjective for every $U \in \mathring{\gamma}(X)$.

Proof. Obviously we may assume that X is Lindelöf. Then we may assume that $J = \mathbb{N}$. Let $U \in \mathring{\mathcal{T}}(X)$ and $s \in \Gamma(U,\mathcal{F})$. We have to extend s to a global section. Let $U_0 := U, s_0 := s$ and $V_n := \cup(U_j \mid 0 \leq j \leq n)$. Successively we construct sections $s_n \in \Gamma(V_n,\mathcal{F})$ with $s_n \mid V_{n-1} = s_{n-1}$. Then these sections s_n glue together to form a section $t \in \Gamma(X,\mathcal{F})$ with $t \mid U = s$. Suppose $n \geq 1$ and s_j is already defined for $j < n$. Since $U_n \cap V_{n-1} \in \mathring{\mathcal{T}}(U_n)$ and $\mathcal{F} \mid U_n$ is sa-flabby, we may extend $s_{n-1} \mid U_n \cap V_{n-1}$ to a section $t_n \in \Gamma(U_n,\mathcal{F})$. Now s_{n-1} and t_n fit together and give the desired s_n.

Proposition 4.8. Let Φ be paracompactifying and $A \subset B \subset X$. Assume that $C \cap A$ is partially quasicompact for every $C \in \Phi$. Let $\mathcal{F}$ be a sa-flabby sheaf on B. Then $\mathcal{F} \mid A$ is a $(\Phi \cap A)$-soft sheaf.

Proof. Let $C \in \Phi$ and $s \in \Gamma(C \cap A,\mathcal{F})$. By Lemma 3.1 we are able to extend s to an open neighbourhood $U \cap B, U \in \mathring{\mathcal{T}}(X)$, of $C \cap A$ in B and hence to a global section $t \in \Gamma(B,\mathcal{F})$.

Our next goal is to show that in many cases sa-flabby and Φ-soft sheaves are Φ-acyclic.

Theorem 4.9. Let Φ be paracompactifying and $A \subset X$. Assume that $A \cap C$ is partially quasicompact for every $C \in \Phi$. Then every $(\Phi \cap A)$-soft sheaf $\mathcal{F}$ on A is $(\Phi \cap A)$-acyclic.

Theorem 4.10. Assume that X is of type (L). Let $\mathcal{F}$ be a sa-flabby sheaf on X. Then $\mathcal{F}$ is Φ-acyclic for every (locally semialgebraic) family Φ of supports on X.

Before proving these theorems we sketch how Theorem 4.9 may be easily derived from well known results in topological sheaf theory in some special cases (which are certainly the most important ones).
Replacing X by X_Φ if necessary we may assume that X is a regular and taut space (Lemma 3.4). Now consider the following cases:

1) $A \subset X$ is closed or $A \in \mathcal{T}(X)$ is a locally semialgebraic subset (or, even more generally, A is a locally proconstructible subset (cf. [S_1]) of X).

2) $A = X^{\max}$.

In both cases we have the continuous canonical retraction $r := r_A : A \longrightarrow A^{\max}$ (I.4.3). Moreover, $r_*\mathcal{F} = \mathcal{F} \mid A^{\max}$ for every sheaf $\mathcal{F}$ on A and $\Gamma_{\Phi\cap A}(A,\mathcal{F}) = \Gamma_{\Phi\cap A^{\max}}(A^{\max},\mathcal{F})$ (Remark 3.7). Thus r_* is an exact functor and we obtain from the Leray-spectral sequence of r ([B, IV.6.])

Proposition 4.11. In each of the cases 1), 2) the cohomology groups $H^q_{\Phi\cap A}(A,\mathcal{F})$ and $H_{\Phi\cap A^{\max}}(A^{\max},\mathcal{F})$ are canonically isomorphic for any sheaf $\mathcal{F}$ on A and every paracompactifying family Φ of supports on X.

Now suppose we are in one of the cases 1), 2) and $\mathcal{F}$ is a $(\Phi \cap A)$-soft sheaf on A. From Lemma 4.4 we know that $\mathcal{F} \mid A^{\max}$ is $(\Phi \cap A^{\max})$-soft. Since $\Phi \cap A^{\max}$ is paracompactifying in the classical sense (Prop. 1.6), $\mathcal{F} \mid A^{\max}$ is $(\Phi \cap A^{\max})$-acyclic ([B, II.9.8]). Hence $\mathcal{F}$ is $(\Phi \cap A)$-acyclic.

The proof of both theorems 4.9 and 4.10 runs in the same way as the analogous proofs in classical sheaf theory (cf. [B, II.5.9], [G, II.4.4.3]). The key point is

Proposition 4.12. Let $A \subset X, \Phi$ be a family of supports on X and $0 \longrightarrow \mathcal{F}' \xrightarrow{\alpha} \mathcal{F} \xrightarrow{\beta} \mathcal{F}'' \longrightarrow 0$ be an exact sequence of sheaves on A. Assume that we are in one of the following cases:

1) $A \cap C$ is partially quasicompact for every $C \in \Phi, \Phi$ is paracompactifying and $\mathcal{F}'$ is $(\Phi \cap A)$-soft.
2) $A = X, X$ is of type (L) and $\mathcal{F}'$ is sa-flabby.

Then $0 \longrightarrow \Gamma_{\Phi \cap A}(A, \mathcal{F}') \xrightarrow{\alpha} \Gamma_{\Phi \cap A}(A, \mathcal{F}) \xrightarrow{\beta} \Gamma_{\Phi \cap A}(A, \mathcal{F}'') \longrightarrow 0$ is also exact.

Proof. Case 1: Let $s \in \Gamma_{\Phi \cap A}(A, \mathcal{F}'')$. We must find some $t \in \Gamma_{\Phi \cap A}(A, \mathcal{F})$ with $\beta(t) = s$. We choose $B, C \in \Phi$ such that C is a neighbourhood of B in X and $\operatorname{supp}(s) \subset B \cap A$. By Lemma I.5.5 there is a set $U \in \dot{T}(X)$ with $B \subset U \subset C$. Suppose we find a section $t \in \Gamma(C \cap A, \mathcal{F})$ with $\beta(t) = s \mid C \cap A$. Then $\beta(t \mid (C \setminus U) \cap A) = 0$, hence $t \mid (C \setminus U) \cap A = \alpha(u), u \in \Gamma((C \setminus U) \cap A, \mathcal{F}')$. Since $\mathcal{F}'$ is $(\Phi \cap A)$-soft we can extend u to a section $u' \in \Gamma(C \cap A, \mathcal{F}')$. Then $t' := t - \alpha(u')$ is 0 on $(C \setminus U) \cap A$ and hence can be extended, by 0, to a section $t_1 \in \Gamma(A, \mathcal{F})$ with $\operatorname{supp}(t_1) \subset C \cap A \in \Phi \cap A$ and $\beta(t_1) = s$. We see that we may assume in the proof that X is regular and paracompact, A is partially quasicompact and $\Phi = cld(X)$.
Let $s \in \Gamma(A, \mathcal{F}'')$. Let $(X_i \mid i \in I)$ be a locally finite affine open cover of X. Then we choose a covering $(B_i \mid i \in I)$ of X by sets $B_i \in \bar{\gamma}(X)$ with $B_i \subset X_i$ for every $i \in I$ (I.5.1). Since $B_i \cap A$ is quasicompact we find, for every $i \in I$, finitely many open semialgebraic subsets U_{ij} of $X_i \quad (j \in J(i))$ which cover $B_i \cap A$ and sections $s_{ij} \in \Gamma(U_{ij} \cap A, \mathcal{F})$ with $\beta(s_{ij}) = s \mid U_{ij} \cap A$. Since $(U_{ij} \mid i \in I, j \in J(i))$ is a locally finite family in X, its union U is an element of $\dot{T}(X)$. Now we take a covering $(B_{ij} \mid i \in I, j \in J(i))$ of U by sets $B_{ij} \in \bar{\gamma}(U)$ with $B_{ij} \subset U_{ij}$ (for all i and j). Let $I_1 := \cup(\{i\} \times J(i) \mid i \in I)$. For every $J \subset I_1$ let $F_J := \cup(B_{ij} \cap A \mid (i,j) \in J)$. Then $F_J \in \bar{T}(U) \cap A$ and $F_{I_1} = A$. We consider pairs (J, t) consisting of a subset J of I_1 and a section $t \in \Gamma(F_J, \mathcal{F})$ with $\beta(t) = s \mid F_J$. We order the set of these pairs inductively by

$$(J, t) \leq (J', t') \text{ if } J \subset J' \text{ and } t' \mid F_J = t.$$

Then we apply Zorn's Lemma and choose a maximal element (J, t). It suffices to prove that $J = I_1$. Assume that there is some $(i,j) \in I_1 \setminus J$. Then $(s_{ij} \mid B_{ij} \cap F_J - t \mid B_{ij} \cap F_J)$ is mapped to zero by β, hence it is the image $\alpha(u)$ of some section $u \in \Gamma(B_{ij} \cap F_J, \mathcal{F}')$. Since $\mathcal{F}'$ is soft, we can extend u to a section $u' \in \Gamma(B_{ij} \cap A, \mathcal{F}')$. Now $s_{ij} \mid (B_{ij} \cap A) - \alpha(u')$ and t glue together to form a section $t' \in \Gamma(F_J \cup (B_{ij} \cap A), \mathcal{F})$ with $\beta(t') = s \mid F_J \cup (B_{ij} \cap A)$. This contradicts the maximality of (J, t). Thus $J = I_1$. Case 1) is proved.
Case 2: We may assume that X is Lindelöf. Let $s \in \Gamma_\Phi(X, \mathcal{F}'')$ and $B \in \Phi \cap T(X)$ with $\operatorname{supp}(s) \subset B$. There is a covering $(U_i \mid i \in I)$ of X by open semialgebraic subsets $U_i \in \mathring{\gamma}(X)$ and sections $s_i \in \Gamma(U_i, \mathcal{F})$ with $\beta(s_i) = s \mid U_i$ for every $i \in I$. Since $\beta(s_i \mid U_i \setminus B) = 0$, there is some $u_i \in \Gamma(U_i \setminus B, \mathcal{F}')$ with $\alpha(u_i) = s_i \mid U_i \setminus B$. Since $U_i \setminus B \in \dot{T}(X)$ and $\mathcal{F}'$ is sa-flabby we can extend u_i to $u_i' \in \Gamma(U_i, \mathcal{F}')$. Replacing s_i by $s_i - \alpha(u_i')$ we may now assume that $s_i \mid U_i \setminus B = 0$. We may also assume that $I = \mathbb{N}$ because X is Lindelöf.
Let $V_n = \cup(U_i \mid 1 \leq i \leq n)$. Successively we construct sections $t_n \in \Gamma(V_n, \mathcal{F})$ such that $t_n \mid V_{n-1} = t_{n-1}, t_n \mid V_n \setminus B = 0$ and $\beta(t_n) = s \mid V_n$. Then these sections t_n glue together to form a section $t \in \Gamma_\Phi(X, \mathcal{F})$ with $\beta(t) = s$. Suppose t_j is already

defined for $j < n (n \geq 1)$. Then $\beta(t_{n-1} \mid V_{n-1} \cap U_n - s_n \mid V_{n-1} \cap U_n) = 0$ and thus $t_{n-1} \mid V_{n-1} \cap U_n - s_n \mid V_{n-1} \cap U_n = \alpha(u), u \in \Gamma(V_{n-1} \cap U_n, \mathcal{F}')$. Necessarily we have $u \mid (V_{n-1} \cap U_n) \setminus B = 0$. By 0 we extend u to a section $u_1 \in \Gamma((V_{n-1} \cap U_n) \cup (U_n \setminus B), \mathcal{F}')$. Then we extend u_1 to a section $u_2 \in \Gamma(U_n, \mathcal{F}')$. This is possible since $\mathcal{F}'$ is sa-flabby. Then t_{n-1} and $s_n + \alpha(u_2)$ glue together to form a section $t_n \in \Gamma(V_n, \mathcal{F})$ with the desired properties. The proof of Proposition 4.12 is finished.

Now we conclude by standard arguments (cf. e.g. [B, II.9]) that soft and sa-flabby sheaves are acyclic.

Corollary 4.13. We consider the situation of Proposition 4.12 and assume in addition that $\mathcal{F}$ is also $(\Phi \cap A)$-soft (case 1) resp. sa-flabby (case 2). Then $\mathcal{F}''$ is $(\Phi \cap A)$-soft resp. sa-flabby.

Proof. First we consider case 1. Let $B \in \Phi$ and $s \in \Gamma(B \cap A, \mathcal{F}'')$. We apply Prop. 4.12 to the partially quasicompact subset $B \cap A$ and obtain a section $t \in \Gamma(B \cap A, \mathcal{F})$ with $\beta(t) = s$. Since $\mathcal{F}$ is $(\Phi \cap A)$-soft we can extend t to a section $t' \in \Gamma(A, \mathcal{F})$. Then $\beta(t') \in \Gamma(A, \mathcal{F}'')$ and $\beta(t') \mid B \cap A = s$.
Case 2 can be handled in the same way.

Now we are able to prove Theorem 4.9. Let $\mathcal{F}$ be a $(\Phi \cap A)$-soft sheaf on A. We choose a flabby resolution $0 \longrightarrow \mathcal{F} \longrightarrow \mathcal{J}^0 \longrightarrow \mathcal{J}^1 \longrightarrow \ldots$ of $\mathcal{F}$. The sheaves $\mathcal{J}^k$ are $(\Phi \cap A)$-soft by Proposition 4.8. Inductively we conclude from Cor. 4.13 that $\ker(\mathcal{J}^k \longrightarrow \mathcal{J}^{k+1})$ is $(\Phi \cap A)$-soft for every $k \geq 0$. Now Proposition 4.12 implies that $0 \longrightarrow \Gamma_{\Phi \cap A}(A, \mathcal{F}) \longrightarrow \Gamma_{\Phi \cap A}(A, \mathcal{J}^0) \longrightarrow \Gamma_{\Phi \cap A}(A, \mathcal{J}^1) \longrightarrow \ldots$ is an exact sequence. This means that $H^q_{\Phi \cap A}(A, \mathcal{F}) = 0$ for $q > 0$.

Theorem 4.10 can be derived in the same way.

Remark 4.14. The statement of Theorem 4.10 remains true without any restriction on X if all members of $\Phi \cap \mathcal{T}(X)$ are assumed to be paracompact. This can be shown by similar arguments.
We will not use this fact in the following.
Sa-flabby sheaves and soft sheaves have a cohomological characterization.

Proposition 4.15. Let $\mathcal{F}$ be a sheaf on X and suppose that X is of type (L). The following statements are equivalent:

a) $\mathcal{F}$ is sa-flabby.
b) $H^q_\Phi(X, \mathcal{F}) = 0$ for $q > 0$ and all (locally semialgebraic) support families Φ.
c) $H^1_A(X, \mathcal{F}) = 0$ for every $A \in \bar{\mathcal{T}}(X)$.

Proof. a) implies b) by Theorem 4.10. We have to prove c) $\Longrightarrow$ a). Let $U \in \mathring{\mathcal{T}}(X)$. Then take a flabby resolution $0 \longrightarrow \mathcal{F} \longrightarrow \mathcal{J}^\bullet$ of $\mathcal{F}$. The sequence $0 \longrightarrow \Gamma_{X \setminus U}(X, \mathcal{J}^\bullet) \longrightarrow \Gamma(X, \mathcal{J}^\bullet) \longrightarrow \Gamma(U, \mathcal{J}^\bullet) \longrightarrow 0$ of complexes is exact and induces the long exact cohomology sequence $0 \longrightarrow \Gamma_{X \setminus U}(X, \mathcal{F}) \longrightarrow \Gamma(X, \mathcal{F}) \longrightarrow \Gamma(U, \mathcal{F}) \longrightarrow H^1_{X \setminus U}(X, \mathcal{F}) \longrightarrow \ldots$. We see that $\Gamma(X, \mathcal{F}) \longrightarrow \Gamma(U, \mathcal{F})$ is surjective.

Notation. Let Y be a topological space and $A \subset Y$ be a locally closed subset. We choose an open subset U of Y such that A is closed in U. Let $i : A \hookrightarrow U$ and $j : U \hookrightarrow X$ be the inclusion maps. For any sheaf $\mathcal{G}$ on U let $j_!\mathcal{G}$ be the extension by zero, i.e. $j_!\mathcal{G}$ is the associated sheaf on Y of the presheaf

$$V \longmapsto \begin{cases} \Gamma(V,\mathcal{G}) & \text{if } V \subset U \\ 0 & \text{if } V \not\subset U. \end{cases}$$

If $\mathcal{F}$ is a sheaf on A, then we denote the sheaf $j_!(i_*\mathcal{F})$ by $\mathcal{F}^Y$. Notice that $\mathcal{F}^Y$ does not depend on the choice of the set U. If $\mathcal{G}$ is a sheaf on Y, then $\mathcal{G}_A$ denotes the sheaf $(\mathcal{G} \mid A)^Y$.

Lemma 4.16. Let Φ be a paracompactifying family of supports on the abstract space X and $A \subset X$ be a locally closed subset. Then a sheaf $\mathcal{F}$ on A is $(\Phi \mid A)$-soft if and only if $\mathcal{F}^X$ is Φ-soft.

Proof. The „if" part follows from Prop. 4.1.ii). For the „only if" part, we note that $\Gamma_\Phi(X,\mathcal{F}^X) = \Gamma_{\Phi|A}(A,\mathcal{F})$ ([B, I.6.6]) and similarly, for $F \subset X$ closed, $\Gamma_{\Phi|F}(F,\mathcal{F}^X) = \Gamma_{\Phi|F\cap A}(F \cap A,\mathcal{F})$. Now the result follows from Prop. 4.1.iii).

Proposition 4.17. Let Φ be a paracompactifying family of supports and $\mathcal{F}$ be a sheaf on X. The following statements are equivalent:

a) $\mathcal{F}$ is Φ-soft.
b) $\mathcal{F}_U$ is Φ-acyclic for every $U \in \dot{\mathcal{T}}(X)$.
c) $H^1_\Phi(X,\mathcal{F}_U) = 0$ for every $U \in \dot{\mathcal{T}}(X)$.

Proof. a) $\Longrightarrow$ b) follows from Theorem 4.9, since $\mathcal{F}_U$ is Φ-soft by Lemma 4.16.
b) $\Longrightarrow$ c) is trivial. To prove c) $\Longrightarrow$ a), we consider a set $A \in \Phi \cap \mathcal{T}(X)$. Then $U := X \setminus A \in \dot{\mathcal{T}}(X)$ and we have the exact sequence

$$0 \longrightarrow \mathcal{F}_U \longrightarrow \mathcal{F} \longrightarrow \mathcal{F}_A \longrightarrow 0$$

of sheaves on X ([G, II.4.10]). The associated long exact cohomology sequence

$$\cdots \longrightarrow \Gamma_\Phi(X,\mathcal{F}) \longrightarrow \Gamma_\Phi(X,\mathcal{F}_A) \longrightarrow H^1_\Phi(X,\mathcal{F}_U) \longrightarrow \cdots$$

shows that $\Gamma_\Phi(X,\mathcal{F}) \longrightarrow \Gamma_\Phi(X,\mathcal{F}_A) = \Gamma(A,\mathcal{F})$ is surjective. Hence $\mathcal{F}$ is Φ-soft (Prop. 4.2).

The following lemma turns out to be very useful.

Lemma 4.18. Let $\mathfrak{K}$ be a class of sheaves of Λ-modules on X, satisfying the following properties.

a) If $0 \longrightarrow \mathcal{F}' \longrightarrow \mathcal{F} \longrightarrow \mathcal{F}'' \longrightarrow 0$ is exact with $\mathcal{F}' \in \mathfrak{K}$, then $\mathcal{F} \in \mathfrak{K}$ if and only if $\mathcal{F}'' \in \mathfrak{K}$.
b) If $(\mathcal{F}_\alpha \mid \alpha \in I)$ is an upward-directed family of subsheaves of some sheaf $\mathcal{F}$ with each $\mathcal{F}_\alpha \in \mathfrak{K}$, then $\varinjlim_{\alpha\in I} \mathcal{F}_\alpha \in \mathfrak{K}$.[1]
c) For any ideal I of Λ and any $U \in \dot{\mathcal{T}}(X)$, $I_U \in \mathfrak{K}$. (I_U the extension by zero of the constant sheaf I on U).

Then $\mathfrak{K}$ consists of all sheaves of Λ-modules.

[1] $\varinjlim \mathcal{F}_\alpha$ denotes the direct limit in the category of sheaves.

Note that the only difference to the corresponding lemma in classical sheaf theory ([B, II.15.10]) is that, in c), we only require that $I_U \in \mathfrak{K}$ for every $U \in \mathring{T}(X)$ and not for every open subset U of X. By [B, II.15.10] it suffices to show that I_V is in $\mathfrak{K}$ where V is an arbitrary open subset of X. This follows from b) and c) since $I_V = \varinjlim_{\substack{U \subseteq V \\ U \in \mathring{T}(X)}} I_U$.

Remark 4.19. Suppose X is semialgebraic. Let $(\mathcal{F}_\alpha \mid \alpha \in I)$ be a direct system of sheaves on X. The set X equipped with the family $\mathring{\gamma}(X)$ of open semialgebraic subsets is a restricted topological space (cf. [DK, §7]). This topology is called the semialgebraic topology of X. It is a generalized topology in the sense of Grothendieck (cf. [A]) and the sheaf theory on X coincides with the sheaf theory on X with respect to semialgebraic topology ([A, II.5], cf. also I, §1). The semialgebraic topology is noetherian, i.e. all admissible open sets $U \in \mathring{\gamma}(X)$ are quasicompact. This implies that $\Gamma(U, \varinjlim \mathcal{F}_\alpha) = \varinjlim \Gamma(U, \mathcal{F}_\alpha)$ for every $U \in \mathring{\gamma}(X)$ ([A, II.5.3]).

Lemma 4.20. Assume X is of type (L). Let $(\mathcal{F}_\alpha \mid \alpha \in I)$ be a direct system of *sa*-flabby sheaves on X. Then $\varinjlim \mathcal{F}_\alpha$ is *sa*-flabby.

Proof. By Prop. 4.7 we may assume that X is affine. Now we know from Remark 4.19 that $\Gamma(U, \varinjlim \mathcal{F}_\alpha) = \varinjlim \Gamma(U, \mathcal{F}_\alpha)$ for every $U \in \mathring{\gamma}(X)$. This proves the lemma.

Proposition 4.21. Suppose X is semialgebraic. Let $(\mathcal{F}_\alpha \mid \alpha \in I)$ be a direct system of sheaves on X and Φ be a (semialgebraic) family of supports on X. Then

$$H^q_\Phi(X, \varinjlim \mathcal{F}_\alpha) = \varinjlim H^q_\Phi(X, \mathcal{F}_\alpha).$$

Proof. There exists a direct system of resolutions

$$0 \longrightarrow \mathcal{F}_\alpha \longrightarrow \mathcal{J}^0_\alpha \longrightarrow \mathcal{J}^1_\alpha \longrightarrow \dots (\alpha \in I)$$

of the sheaves $\mathcal{F}_\alpha$ by injective sheaves $\mathcal{J}^k_\alpha$ ([A, II.5]). Injective sheaves are flabby. From Lemma 4.20 we conclude that

$$0 \longrightarrow \varinjlim \mathcal{F}_\alpha \longrightarrow \varinjlim \mathcal{J}^0_\alpha \longrightarrow \varinjlim \mathcal{J}^1_\alpha \longrightarrow \dots$$

is a resolution of $\varinjlim \mathcal{F}_\alpha$ by *sa*-flabby sheaves. Therefore $H^q_\Phi(M, \varinjlim \mathcal{F}_\alpha)$ may be computed by means of this resolution (Theorem 4.10). From Remark 4.19 we know that $\Gamma(U, \varinjlim \mathcal{J}^k_\alpha) = \varinjlim \Gamma(U, \mathcal{J}^k_\alpha)$ for every $U \in \mathring{\gamma}(X)$. This implies that $\Gamma_\Phi(X, \varinjlim \mathcal{J}^k_\alpha) = \varinjlim \Gamma_\Phi(X, \mathcal{J}^k_\alpha)$. Now Proposition 4.21 follows since the homology functor commutes with direct limits ([Sp, 4.17]).

Recall that a sheaf $\mathcal{F}$ (of Λ-modules) is said to be *flat* if all stalks $\mathcal{F}_x$ are flat Λ-modules. A sheaf $\mathcal{F}$ is flat if and only if $\mathcal{G} \longmapsto \mathcal{G} \otimes \mathcal{F}$ is an exact functor. If Λ is a principal ideal domain, then $\mathcal{F}$ is flat if and only if all stalks $\mathcal{F}_x$ (or, equivalently, all modules $\Gamma(U, \mathcal{F}), U$ open in X) are torsion free Λ-modules.

Proposition 4.22. Assume that Λ is a principal ideal domain. Let Φ be a paracompactifying family of supports on X and $\mathcal{F}$ be a flat and Φ-soft sheaf. Then $\mathcal{F} \otimes \mathcal{G}$ is Φ-soft for every sheaf $\mathcal{G}$ on X.

Proof. Let $\mathfrak{K}$ be the set of sheaves $\mathcal{G}$ on X with $\mathcal{F} \otimes \mathcal{G}$ Φ-soft. We will show that $\mathfrak{K}$ has the properties a), b), c) of Lemma 4.18 and hence consists of all sheaves. Since Φ-soft is a local property (Prop. 4.6) and $\Phi \mid U$ is paracompactifying for every $U \in \mathring{\gamma}(X)$ (Lemma 1.8), we may assume that X is an affine space.
a). Let $0 \longrightarrow \mathcal{G}' \longrightarrow \mathcal{G} \longrightarrow \mathcal{G}'' \longrightarrow 0$ be an exact sequence and $\mathcal{G}' \in \mathfrak{K}$. Since $\mathcal{F}$ is flat, the sequence $0 \longrightarrow \mathcal{F} \otimes \mathcal{G}' \longrightarrow \mathcal{F} \otimes \mathcal{G} \longrightarrow \mathcal{F} \otimes \mathcal{G}'' \longrightarrow 0$ is also exact. By Prop. 4.17.b) we know that $H^q_\Phi(X, (\mathcal{F} \otimes \mathcal{G}')_U) = 0$ for $q > 0\,(U \in \mathring{\gamma}(X))$. From the long exact cohomology sequence we conclude that $H^1_\Phi(X, (\mathcal{F} \otimes \mathcal{G})_U) = H^1_\Phi(X, (\mathcal{F} \otimes \mathcal{G}'')_U)$. Hence $\mathcal{G} \in \mathfrak{K}$ if and only if $\mathcal{G}'' \in \mathfrak{K}$ (Prop. 4.17.c)).
b). By Prop. 4.21 we have $H^q_\Phi(X, \varinjlim(\mathcal{F} \otimes \mathcal{G}_\alpha)_U) = \varinjlim H^q_\Phi(X, (\mathcal{F} \otimes \mathcal{G}_\alpha)_U)$. Hence $\varinjlim \mathcal{G}_\alpha \in \mathfrak{K}$ if every $\mathcal{G}_\alpha \in \mathfrak{K}$ (Prop. 4.17).
c). Every ideal I of Λ is isomorphic, as a Λ-module, to Λ. Thus we only have to prove that $\mathcal{F} \otimes \Lambda_U = \mathcal{F}_U$ is Φ-soft. We already know this (Lemma 4.16). q.e.d.

Proposition 4.23. Assume that Λ is a principal ideal domain and X is of type (L). Let $\mathcal{F}$ be a flat *sa*-flabby sheaf on X and $\mathcal{G}$ be a locally constant sheaf. Then $\mathcal{F} \otimes \mathcal{G}$ is *sa*-flabby.

Proof. „*sa*-flabby" is a local property (Prop. 4.7). Hence we may assume that X is affine and $\mathcal{G}$ is constant, $\mathcal{G} = P$, P a Λ-module. Let $0 \longrightarrow P_0 \longrightarrow P_1 \longrightarrow P \longrightarrow 0$ be a free resolution of P. Then $0 \longrightarrow \mathcal{F} \otimes P_0 \longrightarrow \mathcal{F} \otimes P_1 \longrightarrow \mathcal{F} \otimes P \longrightarrow 0$ is exact and we conclude from Cor. 4.13 that it suffices to prove the assertion for P free. So suppose that P is free. If P is finitely generated then $\mathcal{F} \otimes P$ is obviously *sa*-flabby. In general this follows from Lemma 4.20.

The reason why both propositions 4.22 and 4.23 are true is that semialgebraic subsets Y of an abstract space X are quasicompact. The analogue of Lemma 4.20 stated for flabby sheaves is false in general. The analogue of Proposition 4.22 in classical topological sheaf theory is true if X is a locally compact topological space and Φ is the family of compact subsets ([B, II.14.5, 15.1]). This weaker result is sufficient for all needs in the theory of locally compact spaces since „compact-soft" implies „Φ-soft" for every support family Φ which is paracompactifying in the classical sense ([B, II.15.6]). In semialgebraic sheaf theory we need Prop. 4.22 in full generality as the following example indicates.

Example 4.24. There is a sheaf $\mathcal{F}$ on the semialgebraic space $\mathbf{R}^2$ which is c-soft but is not soft. Let $r > 0$, $B_r(0) := \{x \in \mathbf{R}^2 \mid \| x \| < r\}$ and let $\mathcal{F}_r$ be the sheaf which assigns to every open semialgebraic subset U of $\mathbf{R}^2$ the group of all (set-theoretic) maps $f : U \longrightarrow \mathbf{Z}$ which are locally constant on $U \setminus B_r(0)$. Then $\mathcal{F}_r \mid B_r(0)$ is *sa*-flabby and $\mathcal{F}_r \mid \mathbf{R}^2 \setminus \overline{B_r(0)}$ is the constant sheaf $\mathbf{Z}$. For $r < r'$ we have a canonical map $\mathcal{F}_r \longrightarrow \mathcal{F}_{r'}$. Let $\mathcal{F} := \varinjlim \mathcal{F}_r$. Then $\Gamma(\mathbf{R}^2, \mathcal{F}) = \varinjlim \Gamma(\mathbf{R}^2, \mathcal{F}_r)$ (by Remark 4.19). Since every complete (= compact) semialgebraic subset A of $\mathbf{R}^2$ is contained in some $B_r(0)$ and $\mathcal{F} \mid B_r(0) = \mathcal{F}_r \mid B_r(0)$ is *sa*-flabby, it follows from (4.6) and (4.8) that $\mathcal{F}$ is a c-soft sheaf. But $\mathcal{F}$ is not soft. Consider for example the closed semialgebraic subsets $A_1 := \{(x,0) \in \mathbf{R}^2 \mid x \le -1\}$, $A_2 := \{(x,0) \in \mathbf{R}^2 \mid x \ge 1\}$ and $A := A_1 \cup A_2$. The section $s \in \Gamma(A, \mathcal{F})$ which is defined by $s \mid A_1 := -1$ and $s \mid A_2 := 1$ cannot be extended to a global section $t \in \Gamma(\mathbf{R}^2, \mathcal{F})$.

§5 - The cohomology of subspaces

The cohomology of certain subsets A of an abstract space X is expressed in terms of the cohomology of X and in terms of the cohomologies of the open locally semialgebraic neighbourhoods of A in X.
Let X be an abstract space and Φ be a (locally semialgebraic) family of supports on X.

Theorem 5.1. Suppose either that Φ is paracompactifying and that A is a locally closed subset of X or that Φ is arbitrary and A is closed in X. Let $\mathcal{F}$ be a sheaf on A. Then there is a natural isomorphism.

$$H^*_\Phi(X, \mathcal{F}^X) \cong H^*_{\Phi|A}(A, \mathcal{F}).$$

Proof. The case „A closed in X" is covered by [B, II.10.1]. If A is closed in an open subset U of X, then $\Phi \mid U$ is paracompactifying (Lemma 1.8) and $\Phi \mid A = (\Phi \mid U) \cap A$. By Theorem 4.9 we may compute $H^*_{\Phi|A}(A, \mathcal{F})$ by means of a $(\Phi \mid A)$-soft resolution $0 \longrightarrow \mathcal{F} \longrightarrow \mathcal{J}^0 \longrightarrow \mathcal{J}^1 \longrightarrow \dots$ of $\mathcal{F}$ on A. We know from Lemma 4.16 that $(\mathcal{J}^k)^X$ is Φ-soft for every k. Moreover, $\mathcal{G} \longmapsto \mathcal{G}^X$ is an exact functor. Thus $0 \longrightarrow \mathcal{F}^X \longrightarrow (\mathcal{J}^0)^X \longrightarrow (\mathcal{J}^1)^X \longrightarrow \dots$ is a Φ-soft resolution of $\mathcal{F}^X$ which may be used to compute $H^*_\Phi(X, \mathcal{F})$ (Theorem 4.9). Since $\Gamma_\Phi(X, \mathcal{G}^X) = \Gamma_{\Phi|A}(A, \mathcal{G})$ for every sheaf $\mathcal{G}$ on A ([B, I.6.6]) we obtain the desired result.

Theorem 5.2. Assume Φ is paracompactifying. Let $A \subset B$ be subsets of X and $\mathcal{F}$ be a sheaf on B. Assume that $C \cap A$ is partially quasicompact for each member C of Φ. Then the canonical homomorphism

$$\varinjlim_{A \subset U \in \mathring{\mathcal{T}}(X)} H^q_{\Phi \cap U \cap B}(U \cap B, \mathcal{F}) \longrightarrow H^q_{\Phi \cap A}(A, \mathcal{F})$$

is an isomorphism for every $q \geq 0$.

Proof. The statement is true for $q = 0$ by Proposition 3.2. Let $0 \longrightarrow \mathcal{F} \longrightarrow \mathcal{J}^0 \longrightarrow \mathcal{J}^1 \longrightarrow \dots$ be a resolution of $\mathcal{F}$ by flabby sheaves on B. Then $\mathcal{J}^k \mid U \cap B$ is also flabby for every $U \in \mathring{\gamma}(X)$. Hence $H^*_{\Phi \cap U \cap B}(U \cap B, \mathcal{F})$ is the cohomology of the complex

$$\Gamma_{\Phi \cap U \cap B}(U \cap B, \mathcal{J}^\bullet).$$

The restriction $\mathcal{J}^k \mid A$ of $\mathcal{J}^k$ to A is $(\Phi \cap A)$-soft (Prop. 4.8). In particular $\mathcal{J}^k \mid A$ is $(\Phi \cap A)$-acyclic (Theorem 4.9). Thus $H^*_{\Phi \cap A}(A, \mathcal{F})$ is the cohomology of the complex

$$\Gamma_{\Phi \cap A}(A, \mathcal{J}^\bullet).$$

From the case $q = 0$ we conclude that

$$\varinjlim_{A \subset U \in \mathring{\mathcal{T}}(X)} \Gamma_{\Phi \cap U \cap B}(U \cap B, \mathcal{J}^k) \xrightarrow{\sim} \Gamma_{\Phi \cap A}(A, \mathcal{J}^k).$$

Now Theorem 5.2 follows from the fact that the homology functor commutes with direct limits of chain complexes.

We point out some important special cases of Theorem 5.2.

Corollary 5.3. Assume Φ is paracompactifying and $A \subset X$ is closed. Let $\mathcal{F}$ be a sheaf on A and $\mathcal{G}$ be a sheaf on $A^{\max}$. Let $i: A^{\max} \longrightarrow A$ be the inclusion map. Then we have the canonical isomorphisms

$$H^*_{\Phi \cap A}(A, \mathcal{F}) \cong H^*_{\Phi \cap A^{\max}}(A^{\max}, \mathcal{F})$$
$$H^*_{\Phi \cap A}(A, i_*\mathcal{G}) \cong H^*_{\Phi \cap A^{\max}}(A^{\max}, \mathcal{G}).$$

Proof. Since $i_*\mathcal{G} \mid A^{\max} = \mathcal{G}$, the second statement follows from the first. Replacing X by X_Φ we may assume that X is taut and hence maximally closed. $A^{\max}$ is a partially quasicompact subset of X and every neighbourhood $U \in \mathring{\mathcal{T}}(X)$ of $A^{\max}$ in X contains A. Now we apply Theorem 5.2 with $B := A$ and $A := A^{\max}$.

The statement of Cor. 5.3 is not true in general if Φ is not paracompactifying (cf. Example 3.6).

Corollary 5.4. Let M be a geometric space over R, $\mathcal{F}$ be a sheaf on M and Φ be a paracompactifying support family consisting of partially complete sets. We denote the set M equipped with its strong topology by M_{top} (cf. I.5.4.d, note that M_{top} is a topological subspace of $\tilde{M}$). Let $\mathcal{G}$ be a sheaf on M_{top} and $i: M_{top} \longrightarrow \tilde{M}$ be the inclusion map. Then we have canonical isomorphisms

$$H^*_\Phi(M, \mathcal{F}) \cong H^*_{\Phi_{top}}(M_{top}, \tilde{\mathcal{F}} \mid M_{top})$$
$$H^*_\Phi(M, i_*\mathcal{G}) \cong H^*_{\Phi_{top}}(M_{top}, \mathcal{G}).$$

(Here Φ_{top} denotes the family $\tilde{\Phi} \cap M_{top}$ of supports on M_{top}).

Proof. Since $\mathcal{G} = i^* i_* \mathcal{G}$ the second isomorphism is a consequence of the first. Now $\tilde{C} \cap M_{top} = C_{top}$ is a partially quasicompact subset of $\tilde{M}$ for every $C \in \Phi$ (Ex. I.5.4.d). The only neighbourhood $U \in \mathring{\mathcal{T}}(M)$ of M_{top} in $\tilde{M}$ is $\tilde{M}$. We can apply Theorem 5.2 with $B := \tilde{M}$ and $A := M_{top}$.

Example 5.5. Let M be a locally complete space over R. Then the family $c = c(M)$ of complete semialgebraic subsets of M is paracompactifying (cf. Example 1.7 c) and c_{top} is the family of compact subsets of M_{top} (cf. [DK, §9]). We have

$$H^*_c(M, \mathcal{F}) = H^*_{c_{top}}(M_{top}, \mathcal{F} \mid M_{top})$$

for every sheaf $\mathcal{F}$ on M.

Examples 5.6. a) The hypothesis that all members of Φ are partially complete is necessary in Cor. 5.4 (but see Thm. 5.7 below). Consider e.g. the semialgebraic space $M := R$ over R. Let $\mathcal{F}$ be the sheaf $\mathcal{O}_M$ of semialgebraic functions on M. Then

$$\Gamma(M, \mathcal{O}_M) \subsetneq \Gamma(M_{top}, \mathcal{O}_M \mid M_{top}).$$

For example, the function $f : M \longrightarrow \mathbb{R}$ with graph

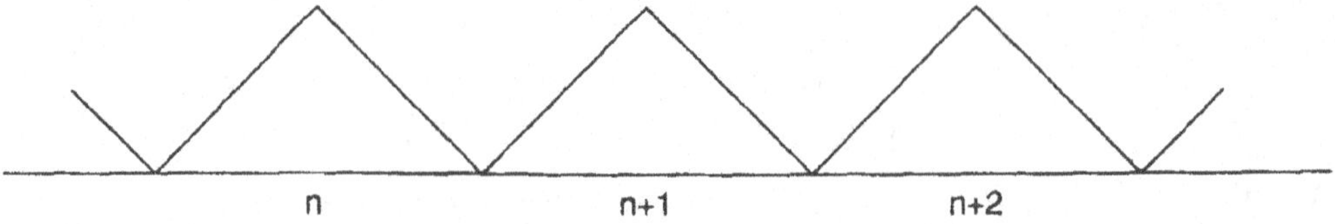

belongs to $\Gamma(M_{top}, \mathcal{O}_M \mid M_{top})$, but clearly the graph of f is not semialgebraic.

Now we describe another counterexample where $\mathcal{F}$ is the direct image $i_*\mathcal{G}$ of a sheaf $\mathcal{G}$ on M_{top} under the inclusion map $i : M_{top} \longrightarrow \tilde{M}$. Consider again $M := \mathbb{R}$. Let A be the closed subset $\cup([2n, 2n+1] \mid n \in \mathbb{Z})$ of M_{top} and $\mathcal{H}$ be the direct image $j_*\mathbb{Z}_A$ of the constant sheaf $\mathbb{Z}$ on A under the inclusion $j : A \subset M_{top}$. Define $\mathcal{G}$ by the exact sequence

$$0 \longrightarrow \mathcal{G} \longrightarrow \mathbb{Z}_{M_{top}} \longrightarrow \mathcal{H} \longrightarrow 0 \quad (*)$$

where $\mathbb{Z}_{M_{top}}$ denotes the constant sheaf $\mathbb{Z}$ on M_{top}. From the long exact cohomology sequence of (*) we immediately conclude that

$$H^1(U_{top}, \mathcal{G}) \cong \Gamma(U_{top}, \mathcal{H})/\Gamma(U_{top}, \mathbb{Z}_{M_{top}}) = \Gamma(U_{top}, \mathcal{H})/\mathbb{Z}$$

for every (bounded or not bounded) open interval U in $\mathbb{R}$. Let $s \in \Gamma(M_{top}, \mathcal{H})$ be the global section of $\mathcal{H}$ with $s \mid [2n, 2n+1] = n$ for every $n \in \mathbb{Z}$. It yields an element $[s] \in H^1(M_{top}, \mathcal{G})$. Since $[s] \mid U_{top} \in H^1(U_{top}, \mathcal{G})$ is not zero for every unbounded open interval $U \subset \mathbb{R}$, $[s]$ yields a non trivial global section $t \in \Gamma(M, R^1 i_*\mathcal{G})$.
Now we consider the exact sequence

$$0 \longrightarrow E_2^{1,0} \longrightarrow E^1 \longrightarrow E_2^{0,1} \longrightarrow E_2^{2,0} \longrightarrow E^2$$

of terms of low degree in the Leray spectral sequence $E_2^{p,q} = H^p(M, R^q i_*\mathcal{G}) \Longrightarrow E^{p+q} = H^{p+q}(M_{top}, \mathcal{G})$ ([CE, Ch. XV, §5]). It says that $0 \longrightarrow H^1(M, i_*\mathcal{G}) \longrightarrow H^1(M_{top}, \mathcal{G}) \longrightarrow \Gamma(M, R^1 i_*\mathcal{G}) \longrightarrow H^2(M, \mathcal{G}) = 0$ is an exact sequence. Since $\Gamma(M, R^1 i_*\mathcal{G}) \neq 0$ we see that $H^1(M, i_*\mathcal{G}) \underset{\neq}{\subset} H^1(M_{top}, \mathcal{G})$.

b) Let X be an affine space and $j : X^{max} \hookrightarrow \tilde{X}$ be the inclusion map. Although $H^*(X^{max}, \mathcal{G}) = H^*(X, j_*\mathcal{G})$ for every sheaf $\mathcal{G}$ on X^{max} by Cor. 5.3, the functor j_* is not exact in general. Take e.g. the unit circle $S^1 = \{x \in \mathbb{R}^2 \mid \| x \| = 1\}$ and $X = \widetilde{S^1}$. Then $X^{max} = S^1_{top}$. We identify $\mathbb{R}$ with the open subset $S^1 - \{(0,1)\}$ of S^1 (e.g. by stereographic projection). Let $\mathcal{H}$ be the sheaf on $\mathbb{R}_{top}$ considered in a) and extend $\mathcal{H}$ by zero to a sheaf $\mathcal{G}$ on S^1_{top}. Furthermore, let $\mathcal{F}$ be the extension by zero of the constant sheaf $\mathbb{Z}$ on $\mathbb{R}_{top}$. Then the obvious map $\mathcal{F} \longrightarrow \mathcal{G}$ is an epimorphism. But if $x \in X$ is one of the (two) ultrafilters converging to the north pole (0,1), then $(j_*\mathcal{F})_x \longrightarrow (j_*\mathcal{G})_x$ is not surjective. We see that j_* is not right exact.

Theorem 5.7. Let M be a geometric space over $\mathbb{R}$ and $\mathcal{F}$ (resp. $\mathcal{G}$) be a locally constant sheaf on M (resp. M_{top}). We denote the inclusion map $M_{top} \longrightarrow \tilde{M}$ by i. Let Φ be an arbitrary family of supports on M. Then the canonical maps

$$H^*_\Phi(M,\mathcal{F}) \longrightarrow H^*_{\Phi_{top}}(M_{top},\mathcal{F} \mid M_{top})$$
$$H^*_\Phi(M,i_*\mathcal{G}) \longrightarrow H^*_{\Phi_{top}}(M_{top},\mathcal{G})$$

are isomorphisms.

Proof. Since $\mathcal{F} \mid M_{top} = i^*\mathcal{F}$ is locally constant and $\mathcal{F} = i_*i^*\mathcal{F}$, it suffices to prove the second statement. We consider the Leray spectral sequence

$$H^p_\Phi(M, R^q i_*\mathcal{G}) \Longrightarrow H^{p+q}_{\Phi_{top}}(M_{top},\mathcal{G})$$

([B, IV.6.1]). Every $U \in \dot{\mathcal{T}}(M)$ has an admissible covering $(U_i \mid i \in I)$ by open and contractible semialgebraic subsets $U_i \in \dot{\gamma}(M)$ (use the triangulation theorem [DK$_1$, §2]). Since $\mathcal{G} \mid (U_i)_{top}$ is constant, we have $H^q((U_i)_{top},\mathcal{G}) = 0$ for $q > 0$. Hence $R^q i_*\mathcal{G} = 0$ for $q > 0$ and we conclude that the edge homomorphism

$$H^*_\Phi(M,i_*\mathcal{G}) \longrightarrow H^*_{\Phi_{top}}(M_{top},\mathcal{G})$$

is an isomorphism.

§6 - Extension of the base field

Let $S \supset R$ be a real closed field extension of the real closed field R. There is a canonical functor from the category of geometric spaces over R to the category of geometric spaces over S (cf. [DK_3, I.2.10], [DK_1, §4]) called „*base field extension*". For a space M over R we denote the space over S obtained from M by the base field extension by $M(S)$. We prove in this section that the cohomology groups of M with coefficients in an arbitrary sheaf $\mathcal{F}$ do not change when the base field is extended. This is an important result and turns out to be useful in many applications. For an affine semialgebraic space M and a locally constant sheaf $\mathcal{F}$ it was already proven in [D, §9].

We consider a geometric space M over R and its extension $M(S)$. The associated abstract space $\widetilde{M(S)}$ of $M(S)$ is just the fibre product $\tilde{M} \times_{\mathrm{Sper}(R)} \mathrm{Sper}(S)$ of $\tilde{M}$ and $\mathrm{Sper}(S)$ over $\mathrm{Sper}(R)$ in the category of abstract spaces ([S_1, III.2.2]). Hence we have a natural map

$$\pi : \widetilde{M(S)} \longrightarrow \tilde{M},$$

the projection onto the first factor. If M is a semialgebraic subspace of an affine R-variety $V = \mathrm{Spec}\, A$,

$$M = \bigcup_{i=1}^{r} \{x \in V(R) \mid f_i(x) = 0, g_{ij}(x) > 0, j = 1, \ldots, s_i\},$$

then $M(S)$ is the semialgebraic subspace

$$\bigcup_{i=1}^{r} \{x \in V \otimes_R S(S) \mid f_i(x) = 0, g_{ij}(x) > 0, j = 1, \ldots, s_i\}.$$

of $V \otimes_R S = \mathrm{Spec}(A \otimes_R S)$. In this case π is the restriction to $\widetilde{M(S)}$ of the map

$$\widetilde{V \otimes_R S(S)} = \mathrm{Sper}(A \otimes_R S) \longrightarrow \mathrm{Sper}(A) = \widetilde{V(R)}$$

which is induced by the natural map $A \longrightarrow A \otimes_R S$.

Let Φ be a (locally semialgebraic) family of supports on M and $\mathcal{F}$ be a sheaf on M. The family

$$\pi^{-1}(\tilde{\Phi}) = \{A \subset \widetilde{M(S)} \mid A \quad \text{closed and} \quad \pi(A) \subset B \quad \text{for some} \quad B \in \tilde{\Phi}\}$$

is locally semialgebraic and hence $\pi^{-1}(\tilde{\Phi}) = \widetilde{\Phi(S)}$ for some family $\Phi(S)$ of supports on $M(S)$. As usual we do not dinstinguish between Φ and $\tilde{\Phi}$ (resp. $\Phi(S)$ and $\widetilde{\Phi(S)}$) in our notation. Obviously $\Phi(S)$ is the family of supports on $M(S)$, which is generated by the sets $A(S), A \in \Phi$. By $\mathcal{F}(S)$ we denote the sheaf on $M(S)$ with $\widetilde{\mathcal{F}(S)} = \pi^*\tilde{\mathcal{F}}$. We have the canonical homomorphism

$$H^q_\Phi(M, \mathcal{F}) = H^q_\Phi(\tilde{M}, \tilde{\mathcal{F}}) \xrightarrow{\pi^*} H^q_{\Phi(S)}(\widetilde{M(S)}, \widetilde{\mathcal{F}(S)}) = H^q_{\Phi(S)}(M(S), \mathcal{F}(S)).$$

The main result in this section is

Theorem 6.1. The canonical map

$$\pi^*: H^q_\Phi(M,\mathcal{F}) \longrightarrow H^q_{\Phi(S)}(M(S),\mathcal{F}(S))$$

is an isomorphism for every $q \geq 0$.

Note that Theorem 6.1 means in particular that $\mathcal{F}(S)$ has the same global sections as $\mathcal{F}$.

Corollary 6.2. The natural map

$$\Gamma_\Phi(M,\mathcal{F}) \longrightarrow \Gamma_{\Phi(S)}(M(S),\mathcal{F}(S))$$

is an isomorphism.

To prove Theorem 6.1 we consider the Leray spectral sequence of π with respect to $\pi^*\tilde{\mathcal{F}} = \widetilde{\mathcal{F}(S)}$ ([B, IV.6], [G, II.4.17]):

$$E_2^{pq} = H^p_\Phi(\tilde{M}, R^q\pi_*(\pi^*\tilde{\mathcal{F}})) \Longrightarrow E^{p+q} = H^{p+q}_{\Phi(S)}(\widetilde{M(S)},\widetilde{\mathcal{F}(S)}).$$

Recall that $R^q\pi_*\mathcal{G}$ ($\mathcal{G}$ a sheaf on $M(S)$) is the sheaf on $\tilde{M}$ which is associated to the presheaf $U \mapsto H^q(\pi^{-1}(U),\mathcal{G})$.

Theorem 6.1 is a consequence of the both following results.

Proposition 6.3. $R^q\pi_*(\pi^*\tilde{\mathcal{F}}) = 0$ for every $q > 0$.

Proposition 6.4. The canonical adjunction homomorphism

$$\alpha: \tilde{\mathcal{F}} \longrightarrow \pi_*\pi^*\tilde{\mathcal{F}}$$

is an isomorphism.

Proposition 6.3 implies that the Leray sequence splits and hence the edge homomorphism $\varepsilon: H^p_\Phi(\tilde{M},\pi_*\pi^*\tilde{\mathcal{F}}) \longrightarrow H^p_{\Phi(S)}(\widetilde{M(S)},\widetilde{\mathcal{F}(S)})$ is an isomorphism. From Proposition 6.4 we conclude that the canonical map

$$\eta: H^p_\Phi(\tilde{M},\tilde{\mathcal{F}}) \longrightarrow H^p_\Phi(\tilde{M},\pi_*\pi^*\tilde{\mathcal{F}})$$

is an isomorphism. Since π^* is the composition $\varepsilon\circ\eta$ of ε and η, Theorem 6.1 follows. Therefore we only have to prove Prop. 6.3 and Prop. 6.4.

Proof of Proposition 6.3. We may assume that M is an affine semialgebraic space over R. Then $M(S)$ is an affine semialgebraic space over S. Let $\mathfrak{K}$ be the class of all sheaves $\mathcal{F}$ on $\tilde{M}$ such that $R^q\pi_*\pi^*\mathcal{F} = 0$ for every $q > 0$. It suffices to show that $\mathfrak{K}$ satisfies the three conditions in Lemma 4.18.

a): Let $0 \longrightarrow \mathcal{F}' \longrightarrow \mathcal{F} \longrightarrow \mathcal{F}'' \longrightarrow 0$ be an exact sequence of sheaves on $\tilde{M}$ with $\mathcal{F}' \in \mathfrak{K}$. Then $0 \longrightarrow \pi^*\mathcal{F}' \longrightarrow \pi^*\mathcal{F} \longrightarrow \pi^*\mathcal{F}'' \longrightarrow 0$ is also an exact sequence. We conclude from the long exact sequence

$$\ldots \to R^q\pi_*(\pi^*\mathcal{F}') \to R^q\pi_*(\pi^*\mathcal{F}) \to R^q\pi_*(\pi^*\mathcal{F}'') \to R^{q+1}\pi_*(\pi^*\mathcal{F}') \to \ldots$$

that $\mathcal{F} \in \mathfrak{K}$ if and only if $\mathcal{F}'' \in \mathfrak{K}$.

b): Let $(\mathcal{F}_\alpha \mid \alpha \in I)$ be a direct system of sheaves with $\mathcal{F}_\alpha \in \mathfrak{K}$ for every $\alpha \in I$. Let

$\mathcal{F} := \varinjlim \mathcal{F}_\alpha$. For every $U \in \mathring{\gamma}(M)$ $\pi^{-1}(\tilde{U}) = \widetilde{U(S)}$ is an open semialgebraic subset of $\widetilde{M(S)}$. Therefore we have

$$H^q(\pi^{-1}(\tilde{U}), \pi^*\mathcal{F}) = \varinjlim H^q(\pi^{-1}(\tilde{U}), \pi^*\mathcal{F}_\alpha)$$

(Prop. 4.21). This means that $R^q\pi_*(\pi^*\mathcal{F}) = \varinjlim R^q\pi_*(\pi^*\mathcal{F}_\alpha)$ and we see that $\mathcal{F} \in \mathfrak{K}$.
c): Let $I \subset \Lambda$ be an ideal and $U \in \mathring{\gamma}(M)$. We have to prove that $I_{\tilde{U}} \in \mathfrak{K}$. Since $\pi^* I_{\tilde{U}}$ is the associated sheaf of the presheaf

$$V \longmapsto \begin{cases} I & \text{if } V \subset \pi^{-1}(\tilde{U}) \\ 0 & \text{else} \end{cases}$$

and $\pi^{-1}(\tilde{U}) = \widetilde{U(S)}$, we see that $\pi^* I_{\tilde{U}} = I_{\widetilde{U(S)}}$. It follows from the definitions (cf. [DK₁, §3], [D₂, §4]) that, for any $W \in \mathring{\gamma}(M)$, we have $H^q(\pi^{-1}(\tilde{W}), I_{\widetilde{U(S)}}) = H^q(W(S), I_{U(S)}) = H^q(W(S), (A \cap W)(S); I)$ with $A := M \setminus U$. The homotopy invariance of semialgebraic cohomology with constant coefficients (cf. §2, [D], [D₂], [DK₁]) implies that $H^q(W(S), (W \cap A)(S); I) = 0$ for $q > 0$ if the pair $(W, W \cap A)$ and therefore, by Tarski's principle, $(W(S), (W \cap A)(S))$ is semialgebraically contractible. We know from the triangulation theorem ([DK₃, 2.13], [DK₁, §2]) that every $x \in \tilde{M}$ has a fundamental system of open neighbourhoods consisting of sets $\tilde{W}$ such that $W \in \mathring{\gamma}(M)$ and $(W, W \cap A)$ is contractible. We see that the stalk $R^q\pi_*(\pi^* I_{\tilde{U}})_x$ is 0 for every $x \in \tilde{M}$ and $q > 0$ and hence $R^q\pi_*(\pi^* I_{\tilde{U}}) = 0$ for $q > 0$. This means that $I_{\tilde{U}} \in \mathfrak{K}$.
The proof of Proposition 6.3 is finished.

Proof of Proposition 6.4. Again we may assume that M is an affine semialgebraic space. By definition $\pi^*\tilde{\mathcal{F}} = \widetilde{\mathcal{F}(S)}$ and $\mathcal{F}(S)$ is the associated sheaf on $M(S)$ of the presheaf $\mathcal{G}$ defined by

$$\Gamma(V, \mathcal{G}) = \varinjlim_{\substack{V \subset U(S) \\ U \in \mathring{\gamma}(M)}} \Gamma(U, \mathcal{F})$$

for $V \in \mathring{\gamma}(M(S))$. We have to prove that the induced map in the stalk

$$\alpha_x : \tilde{\mathcal{F}}_x \longrightarrow \pi_*(\pi^*\tilde{\mathcal{F}})_x$$

is an isomorphism for every $x \in \tilde{M}$. These stalks may be described as follows:

$$\tilde{\mathcal{F}}_x = \varinjlim_{\substack{x \in \tilde{U} \\ U \in \mathring{\gamma}(M)}} \Gamma(U, \mathcal{F}),$$

$$\pi_*(\pi^*\tilde{\mathcal{F}})_x = \varinjlim_{\substack{x \in \tilde{U} \\ U \in \mathring{\gamma}(M)}} \Gamma(U(S), \mathcal{F}(S)).$$

We have

$$\Gamma(U(S), \mathcal{F}(S)) = \varinjlim H^0((V_i \mid i \in I), \mathcal{G}).$$

The direct limit is taken over the set of finite coverings $(V_i \mid i \in I)$ of $U(S)$ by open semialgebraic subsets V_i of $U(S)$. Recall that the Cech-module $H^0((V_i \mid i \in I), \mathcal{G})$ is the

Λ-module of families $(s_i \mid i \in I)$ of sections $s_i \in \Gamma(V_i, \mathcal{G})$ with $s_i \mid V_i \cap V_j = s_j \mid V_i \cap V_j$ (for all $i, j \in I$). We call these families „compatible families".
It is evident that α_x is injective. So it remains to prove that α_x is surjective. Let $s \in \pi_*(\pi^*\tilde{\mathcal{F}})_x$ be represented by a compatible family $(s_i \mid i \in I), s_i \in \Gamma(V_i, \mathcal{G})$, where $(V_i \mid i \in I)$ is a finite covering of $U(S)$ by open semialgebraic subsets and $U \in \mathring{\gamma}(M)$ with $x \in \tilde{U}$. We can choose open semialgebraic subsets $U_i \in \mathring{\gamma}(U)$ with $V_i \subset U_i(S)$ and representatives $t_i \in \Gamma(U_i, \mathcal{F})$ of $s_i (i \in I)$. Since $s_i \mid V_i \cap V_j = s_j \mid V_i \cap V_j$ in $\Gamma(V_i \cap V_j, \mathcal{G})$ there are sets $U_{ij} \in \mathring{\gamma}(M)$ with $V_i \cap V_j \subset U_{ij}(S), U_{ij} \subset U_i \cap U_j$ and $t_i \mid U_{ij} = t_j \mid U_{ij}$ in $\Gamma(U_{ij}, \mathcal{F})$ (for every pair $(i,j) \in I \times I$).

Now we choose a simultaneous semialgebraic triangulation

$$\Phi : X \xrightarrow{\sim} U$$

of U and the semialgebraic subsets $U_i, i \in I$ and $U_{ij}, (i,j) \in I \times I$, of U ([DK$_3$, 2.13], [DK$_1$, §2]). Of course we may assume that $U = X$ and Φ is the identity map. The (finite!) set of open simplices of the finite simplical complex X over R is denoted by $\Sigma(X)$. The extension $X(S)$ of X is a simplicial complex over S. The open simplices of $X(S)$ are the extensions $\sigma(S), \sigma \in \Sigma(X)$.
For every $i \in I$ let $St_{X(S)}(V_i)$ be the star neighbourhood of V_i in $X(S)$ ([DK$_3$, II, §7, Def. 1]). It is the union of those open simplices of $X(S)$ whose closure meets V_i. Hence $St_{X(S)}(V_i)$ is the extension $W_i(S)$ of the subcomplex $W_i = \cup(\sigma \in \Sigma(X) \mid \overline{\sigma(S)} \cap V_i \neq \emptyset)$ of X. Clearly W_i is an open semialgebraic subset of X. Since $V_i \subset U_i(S)$ and U_i is an open subcomplex of X, W_i is contained in U_i. The sets $W_i, i \in I$, cover X since $V_i \subset W_i(S)$ and $(V_i \mid i \in I)$ is a covering of $X(S)$.
Let u_i be the section of $\mathcal{G}$ over $W_i(S)$ which is represented by $t_i \in \Gamma(U_i, \mathcal{F})(i \in I)$. We will prove the following claim:
(*) $u_i \mid W_i(S) \cap V_j = s_j \mid W_i(S) \cap V_j$ in $\Gamma(W_i(S) \cap V_j, \mathcal{G})$ for every pair $(i,j) \in I \times I$.

Obviously $W_i(S) \cap V_j$ is contained in the union $\cup(St_X(\sigma)(S) \mid \sigma \in \Sigma(X), \sigma \subset W_i, \sigma(S) \cap V_j \neq \emptyset\}$. We consider an open simplex $\sigma \in \Sigma(X)$ with $\sigma \subset W_i$ and $\sigma(S) \cap V_j \neq \emptyset$. It suffices to show that $u_i \mid V_j \cap St_X(\sigma)(S) = s_j \mid V_j \cap St_X(\sigma)(S)$. We choose points $x \in \overline{\sigma(S)} \cap V_i$ and $y \in \sigma(S) \cap V_j$ and denote the closed line segment from x to y in $\overline{\sigma(S)}$ by $[x,y]$. Since $(V_l \cap [x,y] \mid l \in I)$ is a finite covering of $[x,y]$ by open semialgebraic subsets we find points $x = a_0 < a_1 < \ldots < a_r = y$ on $[x,y]$ (we order $[x,y]$ in the natural way) and elements $l(k) \in I, 0 \leq k \leq r$, such that $l(0) = i, l(r) = j$ and

$$]a_{k-1}, a_{k+1}[\subset V_{l(k)} \text{ for } 1 \leq k \leq r-1$$

$$[a_0, a_1[\subset V_{l(0)}, [a_{r-1}, a_r] \subset V_{l(r)}.$$

{ Here we denote the open (half open) line segment between two points c and d by $]c,d[$ ($[c,d[$)}.
We have $[a_k, a_{k+1}[\subset V_{l(k)} \cap V_{l(k+1)}$ and hence $]a_k, a_{k+1}[\subset U_{l(k),l(k+1)}(S)$ for $0 \leq k \leq r-1$. Since $]a_k, a_{k+1}[\subset \sigma(S)$ and $U_{l(k),l(k+1)}(S)$ is an open subcomplex of $X(S)$ we see that $St_X(\sigma)(S) = St_{X(S)}(\sigma(S))$ is contained in $U_{l(k),l(k+1)}(S)$ for $0 \leq k \leq r-1$. Hence $St_X(\sigma)$ is contained in $U_{l(k),l(k+1)}$ for $0 \leq k \leq r-1$. Now we conclude from the choice of the sets $U_{l,m}$ that $St_X(\sigma) \subset U_{l(k)}$ for $0 \leq k \leq r$ and $t_{l(k)} \mid St_X(\sigma) = t_{l(k+1)} \mid St_X(\sigma)$ in $\Gamma(St_X(\sigma), \mathcal{F})$ for $0 \leq k \leq r-1$. In particular we see that $t_i \mid St_X(\sigma) = t_j \mid St_X(\sigma)$ in $\Gamma(St_X(\sigma), \mathcal{F})$. This proves our claim (*).

Now we take some set W_i with $x \in \tilde{W}_i$. (Recall that the sets W_j cover X). Statement (*) implies that the compatible families $((s_j \mid W_i(S) \cap V_j) \mid j \in I) \in H^0((W_i(S) \cap V_j \mid j \in I), \mathcal{G})$ and $u_i \in \Gamma(W_i(S), \mathcal{G})$ define the same element in $\Gamma(W_i(S), \mathcal{F}(S))$. This means that $\alpha_x((t_i)_x) = s$ where $(t_i)_x$ denotes the element of $\mathcal{F}_x$ which is represented by $t_i \in \Gamma(W_i, \mathcal{F})$. Thus we see that α_x is indeed surjective. Now Proposition 6.4 and Theorem 6.1 are completely proven.

Example 6.5. If $\mathcal{F}$ is a sheaf on $M(S)$, then, in general, $H^q(M, \pi_*\mathcal{F}) := H^q(\tilde{M}, \pi_*\tilde{\mathcal{F}})$ is not equal to $H^q(M(S), \mathcal{F})$. In particular π_* is not an exact functor. Consider e.g. the affine semialgebraic space $M := \mathbf{R}^2$ over $\mathbf{R}$. Let $S \supset \mathbf{R}$ be a non archimedian real closed field extension of $\mathbf{R}$ and let $\varepsilon > 0$ be an element of S which is infinitesimal over $\mathbf{R}$. Let A be the sphere $\{x \in S^2 \mid \| x \| = \varepsilon\}$ of radius ε in S^2. Let $i : A \longrightarrow S^2$ be the inclusion map and $\mathcal{F}$ be the direct image $i_*\mathbf{Z}_A$ of the constant sheaf $\mathbf{Z}$ on A under i. Then

$$\Gamma(U, R^1\pi_*\mathcal{F}) = H^1(A, \mathbf{Z}) \cong \mathbf{Z}$$

for a set $U \in \mathring{\gamma}(\mathbf{R}^2)$ with $0 \in U$. The exact sequence

$$0 \longrightarrow E_2^{1,0} \longrightarrow E^1 \longrightarrow E_2^{0,1} \longrightarrow E_2^{2,0} \longrightarrow E^2$$

of terms of low degree in the Leray spectral sequence
$E_2^{p,q} = H^p(\mathbf{R}^2, R^q\pi_*\mathcal{F}) \Longrightarrow H^{p+q}(S^2, \mathcal{F}) = E^{p+q}$ ([CE, ch. XV, §5])
yields the exact sequence
$0 \longrightarrow H^1(\mathbf{R}^2, \pi_*\mathcal{F}) \longrightarrow H^1(S^2, \mathcal{F}) = H^1(A, \mathbf{Z}) \xrightarrow{\Upsilon} \Gamma(\mathbf{R}^2, R^1\pi_*\mathcal{F})$. Since Υ is an isomorphism, we see that

$$H^1(\mathbf{R}^2, \pi_*\mathcal{F}) = 0 \neq \mathbf{Z} \cong H^1(S^2, \mathcal{F}).$$

In the following we will give some applications of Theorem 6.1. Let $X \subset \tilde{M}$ be a locally closed subset and $Y := \pi^{-1}(X)$ be its preimage under π. The inclusions $X \hookrightarrow \tilde{M}$ and $Y \hookrightarrow \widetilde{M(S)}$ are denoted by i and j, the map $Y \longrightarrow X$ obtained from π by restriction is denoted by p. Y is a locally closed subset of $\widetilde{M(S)}$.

$$\begin{array}{ccc} Y & \stackrel{j}{\hookrightarrow} & \widetilde{M(S)} \\ p \downarrow & & \downarrow \pi \\ X & \stackrel{i}{\hookrightarrow} & \tilde{M} \end{array}$$

Let $\mathcal{F}$ be a sheaf on X and Φ be a support family on M.

Lemma 6.6. There is a canonical isomorphism

$$\pi^*(\mathcal{F}^{\tilde{M}}) \cong (p^*\mathcal{F})^{\widetilde{M(S)}}.$$

Proof. We choose some open subset U of $\tilde{M}$ such that X is closed in U. Let $i_1 : X \hookrightarrow U, i_2 : U \hookrightarrow \tilde{M}, j_1 : Y \hookrightarrow \pi^{-1}(U)$ and $j_2 : \pi^{-1}(U) \hookrightarrow \widetilde{M(S)}$ be the inclusion maps, and let $p_1 : \pi^{-1}(U) \longrightarrow U$ be the restriction of π. We have canonical maps

(1) $p_1^*(i_1)_*\mathcal{F} \longrightarrow (j_1)_*p^*\mathcal{F}$

(2) $(j_2)_!p_1^*(i_1)_*\mathcal{F} \longrightarrow \pi^*(i_2)_!(i_1)_*\mathcal{F} = \pi^*(\mathcal{F}^{\tilde{M}})$

Looking at the stalks we see that both morphisms are isomorphisms. Hence

(1') $(j_2)_!p_1^*(i_1)_*\mathcal{F} \longrightarrow (j_2)_!(j_1)_*p^*\mathcal{F} = (p^*\mathcal{F})^{\widetilde{M(S)}}$

is also an isomorphism. Combining (1') and (2) we get the canonical isomorphism $\pi^*(\mathcal{F}^{\tilde{M}}) \cong (p^*\mathcal{F})^{\widetilde{M(S)}}$.

From Theorem 6.1 and Lemma 6.6 we obtain

Corollary 6.7. There is a canonical isomorphism

$$H^q_\Phi(M, \mathcal{F}^{\tilde{M}}) \xrightarrow{\sim} H^q_{\Phi(S)}(M(S), (p^*\mathcal{F})^{\widetilde{M(S)}})$$

for every $q \geq 0$.

Corollary 6.8. Suppose X is closed in $\tilde{M}$. Then

$$H^q_{\Phi\cap X}(X, \mathcal{F}) \longrightarrow H^q_{\Phi(S)\cap Y}(Y, p^*\mathcal{F})$$

is an isomorphism for every $q \geq 0$.

This is an immediate consequence of Corollary 6.7 and Theorem 5.1.

Corollary 6.9. Let $x \in \tilde{M}$ and G be a Λ-module. Then

$$H^q(\pi^{-1}(x), G) \cong \begin{cases} G & \text{if } q = 0 \\ 0 & \text{if } q > 0 \end{cases}$$

In particular $\pi^{-1}(x)$ is a connected subset of $\widetilde{M(S)}$.

Proof. We may assume that M is an affine semialgebraic space over R. Then we find an open semialgebraic subset $U \in \mathring{\gamma}(M)$ of M such that x is a closed point of $\tilde{U}$. (There is a semialgebraic function $f \in \Gamma(M, \mathcal{O}_M)$ with $f(x) \neq 0$ and $f(y) = 0$ for all proper specializations y of x. Take $U = \{y \in M \mid f(y) \neq 0\}$). Replacing M by U we may assume that x is closed in $\tilde{M}$. Now the result follows from Cor. 6.8.

Note that $\pi^{-1}(x)$ may be described as follows. As usual let $k(x)$ be the residue field $\mathcal{O}_{\tilde{M},x}/\mathfrak{m}_{\tilde{M},x}$ in x. Then $\pi^{-1}(x)$ is homeomorphic to $\operatorname{Sper} k(x) \otimes_R S$ (cf. I.2.4 or [CR, 2.5]).

Now we assume that M is a locally complete space. Then the family c of complete semialgebraic subsets of M is paracompactifying. If M is paracompact, then the family pc of partially complete locally semialgebraic subsets is also paracompactifying. In general $c(S)$ (resp. $pc(S)$) is *not* the family of complete semialgebraic (resp. partially complete locally semialgebraic) subsets of $M(S)$. Consider e.g. the semialgebraic space $M :=]0,1[$ over $\mathbf{R}$ and let $S \supset \mathbf{R}$ be a non archimedian real closed field extension. If $\varepsilon > 0$ is an element of S infinitesimal over $\mathbf{R}$, then the complete subset $[\varepsilon, 1-\varepsilon]$ of $M(S)$ is not contained in the extension

$A(S)$ of any complete semialgebraic subset A of M. Thus in general $c(S) \overset{\subset}{\neq} c(M(S))$ and $pc(S) \overset{\subset}{\neq} pc(M(S))$.

Nevertheless we are able to prove

Theorem 6.10. Let $\mathcal{F}$ be a sheaf on M.

i) The canonical homomorphism

$$H_c^*(M,\mathcal{F}) \longrightarrow H_c^*(M(S),\mathcal{F}(S))$$

is an isomorphism.

ii) If M is paracompact then

$$H_{pc}^*(M,\mathcal{F}) \longrightarrow H_{pc}^*(M(S),\mathcal{F}(S))$$

is also an isomorphism.

To prove part ii) we choose a completion $M \overset{i}{\hookrightarrow} \bar{M}$ of M ([DK$_3$, II.2.1]). This is possible since M is regular and paracompact. We consider M as a subset of $\bar{M}$. $\bar{M}$ is a partially complete paracompact regular space over R and M is an open locally semialgebraic subset of $\bar{M}$ since M is locally complete. The same is true for the extensions $\bar{M}(S)$ and $M(S)$, i.e. $\bar{M}(S)$ is partially complete, regular, paracompact ([DK$_3$, Appendix B], [Thm 1.2]), and $M(S) \in \mathring{\mathcal{T}}(\bar{M}(S))$.

Applying Theorem 5.1 to the open locally semialgebraic subset M of $\bar{M}$ (with $\Phi = cld$) we obtain

Lemma 6.11. $H_{pc}^*(M,\mathcal{F})$ is canonically isomorphic to $H^*(\bar{M},\mathcal{F}^{\bar{M}})$.

Now Theorem 6.10 follows easily. We know that $H_{pc}^*(M,\mathcal{F}) = H^*(\bar{M},\mathcal{F}^{\bar{M}})$ and $H_{pc}^*(M(S),\mathcal{F}(S)) = H^*(\bar{M}(S),\mathcal{F}(S)^{\bar{M}(S)})$. Corollary 6.7 says that $H^*(\bar{M},\mathcal{F}^{\bar{M}}) \longrightarrow H^*(\bar{M}(S),\mathcal{F}(S)^{\bar{M}(S)})$ is an isomorphism. Hence $H_{pc}^*(N,\mathcal{F}) \longrightarrow H_{pc}^*(M(S),\mathcal{F}(S))$ is an isomorphism.

It remains to prove part i). Here we may assume that M is semialgebraic since $H_c^q(M,\mathcal{F}) = \varinjlim_{N\in\mathring{\gamma}(M)} H_c^q(N,\mathcal{F})$ and $H_c^q(M(S)) = \varinjlim_{N\in\mathring{\gamma}(M)} H_c^q(N(S),\mathcal{F}(S))$. But $c = pc$ if M is semialgebraic and part i) follows from part ii).

§7 - Base Change Theorems

First we recall the definition of the direct image of a sheaf relative to a support family (cf. [B, IV.3]).
Let $f: X \longrightarrow Y$ be a (locally semialgebraic) map between abstract spaces. Let Φ be a family of supports on X. Then there is a left exact functor f_Φ from the category of sheaves on X to the category of sheaves on Y defined as follows:
$f_\Phi \mathcal{F}$ is the associated sheaf on Y of the presheaf

$$U \longrightarrow \Gamma_{\Phi \cap f^{-1}(U)}(f^{-1}(U), \mathcal{F})$$

($U \subset Y$ open, $\mathcal{F}$ a sheaf on X).
If U is quasicompact (e.g. $U \in \mathring{\gamma}(Y)$) then $\Gamma(U, f_\Phi \mathcal{F}) = \Gamma_{\Phi \cap f^{-1}(U)}(f^{-1}(U), \mathcal{F})$. For $\Phi = cld$ f_Φ is just the usual direct image functor f_*. The right derived functors of f_Φ are denoted by $R^q f_\Phi$.
In this section we want to study the connection between the stalks $(R^q f_\Phi \mathcal{F})_y$ and the cohomologies of the fibres $f^{-1}(y)$.

Theorem 7.1. Let $f: X \longrightarrow Y$ be a map between abstract spaces and Φ be a family of supports on X. Assume that, for every $y \in Y$, the fibre $f^{-1}(y)$ has a neighbourhood $U \in \mathring{\gamma}(X)$ such that $\Phi \cap U$ is paracompactifying. Furthermore, we assume that $f(A)$ is closed in Y for every $A \in \Phi \cap \bar{\mathcal{T}}(X)$. Then the canonical map

$$(R^q f_\Phi \mathcal{F})_y \longrightarrow H^q_{\Phi \cap f^{-1}(y)}(f^{-1}(y), \mathcal{F})$$

is an isomorphism for every $q \geq 0$ and every $y \in Y$.

Before we start to prove Theorem 7.1 we discuss the condition that f is closed on the locally semialgebraic members of Φ.

Proposition 7.2. Let $f: X \longrightarrow Y$ be a map between abstract spaces. Assume that every fibre $f^{-1}(y)$ of f has a fundamental system $\mathfrak{U}$ of open neighbourhoods in X with $\mathfrak{U} \subset \mathring{\mathcal{T}}(X)$. (This condition is satisfied if f is semialgebraic or, more generally, if every fibre $f^{-1}(y)$ has a paracompact neighbourhood $A \in \mathcal{T}(X)$ in X, cf. I.5.5). Then f is a closed map if and only if $f(A)$ is closed in Y for every $A \in \bar{\mathcal{T}}(X)$.

Proof. We only have to prove the „if"-part. So we consider a closed subset B of X and a point $y \in \overline{f(B)}$. Suppose $f^{-1}(y) \cap B$ is empty. Then we find some $U \in \mathring{\mathcal{T}}(X)$ with $f^{-1}(y) \subset U$ and $U \cap B = \emptyset$. Hence $B \subset A := X \setminus U$. Since $f(A)$ is closed we conclude that $y \in f(A)$. This is a contradiction.

We see that our assumption in Theorem 7.1 means that the restriction $f \mid A$ of f to any $A \in \Phi$ is a closed map.

Example 7.3. Let $f: M \longrightarrow N$ be a map between geometric spaces over R and Φ be a paracompactifying family of supports on M. Assume $f(A) \in \bar{\mathcal{T}}(N)$ for every $A \in \Phi$ (e.g. $f \mid A$ proper for every $A \in \Phi$). Then $\tilde{f}(\tilde{A}) = \widetilde{f(A)}$ is closed in $\tilde{N}$ for every $A \in \Phi$. Hence the assumptions of Theorem 7.1 are satisfied for $\tilde{f}$ and Φ.

Example 7.4. Let $f: M \longrightarrow N$ be a map between locally complete geometric spaces over R. Let c be the family of semialgebraic complete subsets of M. Then

$$(R^q f_c \mathcal{F})_y = H^q_{c\cap \tilde{f}^{-1}(y)}(\tilde{f}^{-1}(y), \mathcal{F}) \qquad (*)$$

for every $q \geq 0$ and every $y \in \tilde{N}$.

Remark 7.5. Example 7.4 reminds us of a familiar result in classical topology. Namely, if f were a continuous map between locally compact topological spaces and c were the family of compact subsets of M, then f_c would be the functor „Direct image with proper support" and the equality (*) also holds (cf. e.g. [SHS, exp. 3]). But this analogy to the classical case is purely formal and somewhat misleading.
Consider e.g. the semialgebraic map $f: \mathbf{R}^2 \longrightarrow \mathbf{R}, (x,y) \mapsto x$, and the sheaf $\mathcal{F} = i_* \mathbf{Z}_A$ on $\mathbf{R}^2$ where $A = \{(x,y) \in \mathbf{R}^2 \mid 0 \leq y \leq 1\}$ and $i: A \hookrightarrow \mathbf{R}^2$ is the inclusion map. Then $\Gamma(\mathbf{R}, f_c\mathcal{F}) = \Gamma_c(\mathbf{R}^2, \mathcal{F}) = 0$. But if f_c were the functor $f_!$ (= „Direct image with proper support") then we would expect $\Gamma(\mathbf{R}, f_c\mathcal{F}) = \Gamma(\mathbf{R}^2, \mathcal{F}) = \mathbf{Z}$ since the supports of all sections of $\mathcal{F}$ are proper over $\mathbf{R}$. (The real functor $f_!$ will be studied in the next section).
Moreover, consider a fibre $\tilde{f}^{-1}(y), y \in \tilde{N}$. $\tilde{f}^{-1}(y)$ is a geometric space over the real closed extension $k(y)$ of R. It is locally complete by [DK$_3$, App. B] since $\tilde{f}^{-1}(y)$ may be identified with a fibre of the map $f_{k(y)}: M(k(y)) \longrightarrow N(k(y))$ obtained from f by extension of the base field ([S$_1$, III.2]). Now we observe that in general $c \cap \tilde{f}^{-1}(y)$ is *not* the family of complete semialgebraic subsets of $\tilde{f}^{-1}(y)$. Here is (an almost trivial) example. Let $M =]0,1[$ be the open unit interval over $\mathbf{R}$ and $f: M \longrightarrow M$ be the identity map. Now consider the fibre $\tilde{f}^{-1}(y)$ where $y \in \tilde{M}$ is the ultrafilter converging to 0.

Proof of Theorem 7.1.

Theorem 5.2 says that

(1) $\quad H^q_{\Phi\cap f^{-1}(y)}(f^{-1}(y), \mathcal{F}) = \varinjlim_{f^{-1}(y)\subset U \in \mathring{T}(X)} H^q_{\Phi\cap U}(U, \mathcal{F}).$

The right hand side of this equation is equal to

(2) $\quad \varinjlim_{A\in\Phi} \varinjlim_{f^{-1}(y)\subset U\in \mathring{T}(X)} H^q_{A\cap U}(U, \mathcal{F}).$

We choose some $V \in \mathring{T}(X)$ such that $f^{-1}(y) \subset V$ and $\Phi \cap V$ is paracompactifying. Then we consider some fixed $A \in \Phi$ and take some $B \in \Phi \cap T(X)$ such that $B \cap V$ is a neighbourhood of $A \cap V$ in V. We have

(3) $\quad \varinjlim_{f^{-1}(y)\subset U\in \mathring{T}(X)} H^q_{A\cap U}(U, \mathcal{F}) = \varinjlim_{f^{-1}(y)\subset U\in \mathring{T}(V)} H^q_{A\cap U}(B\cap U, \mathcal{F}).$

Let $U \in \mathring{T}(V)$ with $f^{-1}(y) \subset U$. Since $f \mid B: B \longrightarrow Y$ is a closed map, $f(B \setminus U)$ is a closed subset of Y which does not contain y. Hence we find some $W \in \mathring{T}(Y)$ with $y \in W$ and $f^{-1}(W) \cap B \subset U \cap B$. We conclude that the right hand side of (3) is equal to

(4) $\quad \varinjlim_{y\in W\in \mathring{T}(Y)} H^q_{A\cap f^{-1}(W)}(B \cap f^{-1}(W), \mathcal{F})$

This in turn is equal to

(5) $\quad \varinjlim_{y\in W\in \mathring{T}(Y)} H^q_{A\cap f^{-1}(W)}(f^{-1}(W), \mathcal{F}).$

Combining (5) with (1) and (2) we obtain

$$(6)\quad H^q_{\Phi\cap f^{-1}(y)}(f^{-1}(y),\mathcal{F}) = \varinjlim_{y\in W\in \mathring{T}(Y)} H^q_{\Phi\cap f^{-1}(W)}(f^{-1}(W),\mathcal{F}).$$

This completes the proof of Theorem 7.1 since $R^q f_\Phi \mathcal{F}$ is the associated sheaf of the presheaf

$$W \mapsto H^q_{\Phi\cap f^{-1}(W)}(f^{-1}(W),\mathcal{F})$$

([B, IV.4]).

Remark 7.6. (on proper maps). A locally semialgebraic map $f: M \longrightarrow N$ between separated geometric spaces over R is called proper if it is universally closed in the category of geometric spaces over R (cf. [DK$_3$, I, §5, Def. 2]). A locally semialgebraic map $g: X \longrightarrow Y$ between abstract spaces is called proper if it is separated and universally closed in the category of abstract spaces ([S$_1$,II.6.1]). Schwartz proves that a semialgebraic map $f: M \longrightarrow N$ between (separated) geometric spaces M and N is proper if and only if the map $\tilde{f}: \tilde{M} \longrightarrow \tilde{N}$ between abstract spaces is proper ([S$_1$, III.4.10]).

We consider a map $f: X \longrightarrow Y$ between abstract spaces which is locally of finite type (cf. I, §2, Def. 1). Then the fibre $f^{-1}(y)$ over a point $y \in Y$ is a geometric space over $k(y)$ (I.2.5). Let Φ be a paracompactifying family of supports on X such that $f \mid A: A \longrightarrow Y$ is proper for every $A \in \Phi \cap \bar{T}(X)$. Let $g: Z \longrightarrow Y$ be an arbitrary locally semialgebraic map. We form the cartesian square

$$\begin{array}{ccc} X & \xleftarrow{g'} & X \times_Y Z \\ f\downarrow & & \downarrow f' \\ Y & \xleftarrow[g]{} & Z \end{array}$$

and denote the family $(g')^{-1}\Phi$ by Φ'.

Lemma 7.7. Let $A \in T(X)$ be a regular and paracompact set. Assume Z is affine. Then $A \times_Y Z = (g')^{-1}(A)$ is regular and paracompact.

Proof. We choose a locally finite covering $(A_i \mid i \in I)$ of A by *closed* affine semialgebraic subsets $A_i \in \bar{\gamma}(A)$ such that every A_i is mapped into an open affine semialgebraic subset of Y. This is possible by (I.5.1). Then $A_i \times_Y Z$ is an affine space and $(A_i \times_Y Z \mid i \in I)$ is a locally finite covering of $A \times_Y Z$ by *closed* semialgebraic affine (and hence regular) subsets. Thus $A \times_Y Z$ is regular and paracompact.

Let $\mathcal{F}$ be a sheaf on X. Then there is a canonical base change homomorphism

$$\alpha: g^* R^q f_\Phi \mathcal{F} \longrightarrow R^q f'_{\Phi'}(g'^* \mathcal{F})$$

for every $q \geq 0$.

Theorem 7.8. (Proper base change). α is an isomorphism for every $q \geq 0$.

Proof. First observe that $(g')^{-1}(A) = A \times_Y Z$. Hence $f' \mid (g')^{-1}(A)$ is proper for every $A \in \Phi \cap \bar{T}(X)$. This implies that $f'(B)$ is a closed subset of Z for every $B \in \Phi' \cap \bar{T}(X \times_Y Z)$.

Moreover, if $U \subset Z$ is an affine open subset then $\Phi' \cap (X \times_Y U) = \Phi' \cap (f')^{-1}(U)$ is paracompactifying by Lemma 7.7. We see that f' (and of course also f) satisfies the hypotheses of Theorem 7.1. Let $z \in Z$ and $y = g(z)$. We conclude from Theorem 7.1 that
$(g^* R^q f_\Phi \mathcal{F})_z = (R^q f_\Phi \mathcal{F})_y = H^q_{\Phi \cap f^{-1}}(f^{-1}(y), \mathcal{F})$,
$(R^q f'_{\Phi'}(g'^* \mathcal{F}))_z = H^q_{\Phi' \cap f'^{-1}(z)}(f'^{-1}(z), g'^* \mathcal{F})$.
Now we observe that

$$f'^{-1}(z) = X \times_Y Z \times_Z \operatorname{Sper} k(z) = X \times_Y \operatorname{Sper} k(y) \times_{\operatorname{Sper} k(y)} \operatorname{Sper} k(z)$$

is the geometric space $f^{-1}(y)(k(z))$ over $k(z)$ obtained from the geometric space $f^{-1}(y)$ over $k(y)$ by extension of the base field $k(z) \supset k(y)$. Moreover, we have $g'^* \mathcal{F} \mid f'^{-1}(z) = (\mathcal{F} \mid f^{-1}(y))(k(z))$ (same notation as in §6) and $\Phi' \cap f'^{-1}(z) = (\Phi \cap f^{-1}(y))(k(z))$. Now we conclude from Theorem 6.1 that

$$H^q_{\Phi \cap f^{-1}(y)}(f^{-1}(y), \mathcal{F}) = H^q_{\Phi' \cap f'^{-1}(z)}(f'^{-1}(z), g'^* \mathcal{F}).$$

This shows that α is an isomorphism.

§8 - Direct image with proper support

Let M be a locally complete geometric space over R. Recall that locally complete spaces are regular. Hence every semialgebraic subset A of N is affine (cf. [DK$_3$, I, §7], [Ro]).

Remark 8.1. Let $(M_i \mid i \in I)$ be an admissible covering of M by open semialgebraic subsets. Since M_i is locally complete, the abstract space $\tilde{M}_i$ is isomorphic (as a locally ringed space) to $\operatorname{Spec} \Gamma(M_i, \mathcal{O}_M)$ ([S, §4, Prop. 76]). Hence $\tilde{M}$ is a scheme with affine open cover $(\operatorname{Spec} \Gamma(M_i, \mathcal{O}_M) \mid i \in I)$.
Now let A be a closed subset of $\tilde{M}$. We endow A with the reduced subscheme structure of $\tilde{M}$. Schwartz's work implies that the locally ringed space A we obtain in this way is an abstract semialgebraic space ([S$_1$, I.4.5, II.2]). We see that we can equip every closed subset A of $\tilde{M}$ with a subspace structure in a natural way. The inclusion $i: A \hookrightarrow M$ is a locally semialgebraic map. (More generally Schwartz defines a subspace structure on every locally proconstructible subset, cf. [S$_1$, II.2]).

Now we consider a locally semialgebraic map $f: M \longrightarrow N$ between locally complete spaces over R. Let $\mathcal{F}$ be a sheaf on M. We define a sheaf $f_! \mathcal{F}$ on N by

$$\Gamma(U, f_!\mathcal{F}) := \{s \in \Gamma(f^{-1}(U), \mathcal{F}) \mid \tilde{f} \mid \operatorname{supp}(s): \operatorname{supp}(s) \to \tilde{U} \text{ is a proper and semialgebraic map}\}$$

for $U \in \mathring{\mathcal{T}}(N)$.

This definition makes sense since $\operatorname{supp}(s)$ is closed in $\widetilde{f^{-1}(U)} = \tilde{f}^{-1}(\tilde{U})$ and hence $\tilde{f} \mid \operatorname{supp}(s): \operatorname{supp}(s) \hookrightarrow \tilde{f}^{-1}(\tilde{U}) \xrightarrow{\tilde{f}} \tilde{U}$ is a locally semialgebraic map. Since „proper“ is a property which may be checked locally on the image space $f_!\mathcal{F}$ is a sheaf on N.

Proposition 8.2. Suppose N is paracompact. Let $A \subset \tilde{M}$ be a closed subset such that $\tilde{f} \mid A: A \longrightarrow \tilde{N}$ is proper and semialgebraic. Then there is some set $B \in \bar{\mathcal{T}}(M)$ such that $\tilde{B}$ is a neighbourhood of A in $\tilde{M}$ and $f \mid B: B \longrightarrow N$ is proper and semialgebraic.

Proof. We choose a locally finite covering $(U_i \mid i \in I)$ of N by open semialgebraic subsets. The closures $N_i = \bar{U}_i$ of the sets U_i in N are also semialgebraic. Assume we find sets $B_i \in \bar{\mathcal{T}}(f^{-1}(N_i))$ such that $\tilde{B}_i$ is a neighbourhood of $A \cap \tilde{f}^{-1}(\tilde{N}_i)$ in $\tilde{f}^{-1}(\tilde{N}_i)$ and $f \mid B_i: B_i \longrightarrow N_i$ is proper and semialgebraic. Then $B := \cup(B_i \mid i \in I)$ is a set with the desired properties. Therefore we may assume that N is semialgebraic.
Then A is semialgebraic, hence quasicompact and there is some $U \in \mathring{\gamma}(M)$ with $A \subset \tilde{U}$. Replacing M by U we may also assume that M is semialgebraic. Now we can choose a proper extension

$$\begin{array}{ccc} M & \overset{i}{\hookrightarrow} & P \\ & f \searrow \quad \swarrow \bar{f} & \\ & N & \end{array}$$

of f ([DK$_3$, II, §12, 12.10], i is an open dense embedding, $\bar{f}$ a proper map). We regard i is an inclusion map. Since $\tilde{f} \mid A: A \longrightarrow \tilde{N}$ is proper, the inclusion $A \hookrightarrow \tilde{P}$ is a closed embedding. By (I.5.3) there exist disjoint open semialgebraic subsets U, V of P with $A \subset \tilde{U}$ and $P \setminus M \subset V$. We set $B := P \setminus V$. Then $B \in \bar{\gamma}(M)$, $\tilde{B}$ is a neighbourhood of A in $\tilde{M}$ and $f \mid B: B \longrightarrow N$ is a proper semialgebraic map. q.e.d.

Let Φ be the family of supports on M consisting of those sets $A \in \tilde{T}(M)$ such that $f \mid A : A \longrightarrow N$ is a proper semialgebraic map. (N.B. A proper locally semialgebraic map is semialgebraic if every component of its domain is Lindelöf, cf. [DK_3, I.5.9]).

Corollary 8.3. Let y be a point in $\tilde{N}$. Then $\tilde{f}^{-1}(y)$ is a locally complete geometric space over $k(y)$. If N is paracompact and y is a closed point then $\Phi \cap \tilde{f}^{-1}(y)$ is the family $c(\tilde{f}^{-1}(y))$ of complete semialgebraic subsets of $\tilde{f}^{-1}(y)$.

Proof. The first statement was already proven in Remark 7.5. If y is closed in $\tilde{N}$, then the inclusion $\{y\} = \operatorname{Sper} k(y) \hookrightarrow \tilde{N}$ is a proper map. If $A \in \bar{\gamma}(\tilde{f}^{-1}(y))$ is complete (over $k(y)$), then $\tilde{f} \mid A : A \longrightarrow \operatorname{Sper} k(y) \hookrightarrow \tilde{N}$ is proper (cf. [S_1, II.6]). We conclude from Prop. 8.2 that there is some $B \in \Phi$ with $A \subset \tilde{B} \cap \tilde{f}^{-1}(y)$. Thus $c(\tilde{f}^{-1}(y)) \subset \Phi \cap \tilde{f}^{-1}(y)$. The other inclusion is obvious.

Corollary 8.4. Assume that N is paracompact. Then Φ is paracompactifying.

Proof. Let $A \in \Phi$. Since $f \mid A : A \longrightarrow N$ is semialgebraic, the set A is paracompact. As a locally complete set it is regular. By Proposition 8.2 A has a neighbourhood B in M with $B \in \Phi$.

Lemma 8.5. Suppose N is paracompact. Let $x \in \tilde{N}$ be a closed point and $\mathcal{F}$ be a sheaf on M. Then

$$(f_! \mathcal{F})_x = (f_\Phi \mathcal{F})_x.$$

Proof. Obviously $f_\Phi \mathcal{F}$ is a subsheaf of $f_! \mathcal{F}$ and hence $(f_\Phi \mathcal{F})_x \subset (f_! \mathcal{F})_x$. Now consider some $U \in \mathring{\gamma}(N)$ with $x \in \tilde{U}$ and a section $s \in \Gamma(U, f_! \mathcal{F})$. We have $s \in \Gamma(f^{-1}(U), \mathcal{F})$ and $\tilde{f} \mid \operatorname{supp}(s) : \operatorname{supp}(s) \longrightarrow \tilde{U}$ is proper. By Prop. 8.2 we find some $A \in \bar{\gamma}(f^{-1}(U))$ such that $\operatorname{supp}(s) \subset \tilde{A}$ and $f \mid A : A \longrightarrow U$ is proper. By Proposition I.3.15 there is some $V \in \mathring{\gamma}(N)$ with $x \in \tilde{V}$ and $\bar{V} \subset U$. Then $A \cap f^{-1}(\bar{V}) \in \Phi$ and hence $\operatorname{supp}(s \mid f^{-1}(V)) \in \Phi \cap f^{-1}(V)$. We see that the germ s_x of s is indeed contained in $(f_\Phi \mathcal{F})_x$.

Remark 8.6. If x is not a closed point of $\tilde{N}$, then we may have $(f_\Phi \mathcal{F})_x \subsetneq (f_! \mathcal{F})_x$. Consider e.g. the map $f : \mathbf{R}^2 \longrightarrow \mathbf{R}, (x, y) \mapsto x$, the closed subset $A := \{(x, y) \mid x \cdot y \leq 1\}$ of $\mathbf{R}^2$ and the direct image $\mathcal{F} = i_* \mathbf{Z}_A$ of the constant sheaf $\mathbf{Z}_A$ under $i : A \subset \mathbf{R}^2$. Let $x \in \tilde{\mathbf{R}}$ be the ultrafilter generated by the intervals $]0, \varepsilon[, \varepsilon > 0$. Then $(f_\Phi \mathcal{F})_x = 0$ and $(f_! \mathcal{F})_x = \mathbf{Z}$.

Now let $g : N' \longrightarrow N$ be another locally semialgebraic map with N' locally complete. Consider the cartesian square

$$\begin{array}{ccc} M & \xleftarrow{g'} & M \times_N N' \\ f \downarrow & & \downarrow f' \\ N & \xleftarrow[g]{} & N' \end{array}$$

and let $\mathcal{F}$ be a sheaf on M. There is a canonical base change homomorphism $\alpha : g^* R^q f_! \mathcal{F} \longrightarrow R^q f'_!(g'^* \mathcal{F})$.

Theorem 8.7. α is an isomorphism.

Proof. We have to check this in the stalks. Hence we may assume that N and N' are semialgebraic. Let $x \in \tilde{N}'$, and $y := \tilde{f}(x) \in \tilde{N}$. Replacing N and N' by suitable open semialgebraic subsets we may assume that x and y are closed points of $\tilde{N}'$ and $\tilde{N}$. Let $0 \longrightarrow \mathcal{F} \longrightarrow \mathcal{J}^\bullet$ be an injective resolution. We know from Lemma 8.5 that $(f_!\mathcal{J}^\bullet)_y = (f_\Phi \mathcal{J}^\bullet)_y$. Hence we have

$$(g^* R^q f_! \mathcal{F})_x = (R^q f_! \mathcal{F})_y = (R^q f_\Phi \mathcal{F})_y.$$

Since Φ is paracompactifying (Corollary 8.4) we conclude from Theorem 7.1 and Corollary 8.3 that

$$(g^* R^q f_! \mathcal{F})_x = H_c^q(\tilde{f}^{-1}(y), \mathcal{F}).$$

The same arguments show that $(R^q f'_!(g'^* \mathcal{F}))_x = H_c^q(\tilde{f}'^{-1}(x), g'^* \mathcal{F})$. $\tilde{f}^{-1}(y)$ is a geometric space over $k(y)$ and $\tilde{f}'^{-1}(x)$ is the space $\tilde{f}^{-1}(y)(k(x))$ obtained from $\tilde{f}^{-1}(y)$ by extension of the base field $k(x) \supset k(y)$. Moreover, $g'^* \mathcal{F} \mid \tilde{f}'^{-1}(x) = (\mathcal{F} \mid \tilde{f}^{-1}(y))(k(x))$ (cf. proof of 7.8). Now we conclude from Theorem 6.10 that

$$H_c^q(\tilde{f}^{-1}(y), \mathcal{F}) = H_c^q(\tilde{f}'^{-1}(x), g'^* \mathcal{F})$$

and Theorem 8.7 is proven.

From the proof we obtain

Corollary 8.8. For every $x \in \tilde{N}$ the stalk $(R^q f_! \mathcal{F})_x$ is canonically isomorphic to $H_c^q(\tilde{f}^{-1}(x), \mathcal{F})$.

Lemma 8.9. Let P be a locally complete space over R and $(\mathcal{F}_\alpha \mid \alpha \in I)$ be a direct system of sheaves on P. Then

$$H_c^*(P, \varinjlim \mathcal{F}_\alpha) = \varinjlim H_c^*(P, \mathcal{F}_\alpha).$$

Proof. We may assume that P is semialgebraic (cf. proof of Thm. 6.10). Then the statement follows from Prop. 4.21.

Corollary 8.10. Let $(\mathcal{F}_\alpha \mid \alpha \in I)$ be a direct system of sheaves on M. Then

$$R^q f_!(\varinjlim \mathcal{F}_\alpha) = \varinjlim R^q f_! \mathcal{F}_\alpha$$

for every $q \geq 0$.

This follows from Cor. 8.8 and Lemma 8.9.

Remarks 8.11. i) If M is paracompact and f is a proper map, then Theorem 8.7 is contained in Theorem 7.8. (Take $\Phi = \bar{T}(M)$ there).
ii) It seems that the results of this section suffice to prove a semialgebraic analogue of Verdier duality (by the same proof as in the theory of locally compact spaces cf. [V]).
iii) It is possible to prove the results of this section more generally for a map $f: X \longrightarrow Y$ between abstract spaces which is locally of finite type. But this requires a more detailed study of maps of finite type and completions in the abstract category (cf. [S, §6], [S_1, II.7]). Once this is done the arguments are almost identical with those we used here in the geometric case.

§9 - Cohomological dimension

For later use we prove some results on the cohomological dimension of geometric spaces.

Lemma 9.1. Let M be an affine semialgebraic space over R of dimension n and $V \in \mathring{\gamma}(M)$. Then $H^q(M,\mathcal{F}) = 0$ for every $q > n$ and the natural map

$$H^n(M,\mathcal{F}) \longrightarrow H^n(V,\mathcal{F})$$

is surjective for every sheaf $\mathcal{F}$ on M.

Proof. Let $\mathfrak{U}$ be the set of finite coverings $(U_i \mid i \in I)$ of M by open semialgebraic subsets which have the following properties:
a) $U_{i_0} \cap \ldots \cap U_{i_{n+1}} = \emptyset$ for pairwise different indices $i_0, \ldots, i_{n+1}$.
b) $(U_i \mid i \in I(V))$ is a covering of V where $I(V) = \{i \in I \mid U_i \subset V\}$.

Let $\mathfrak{V}$ be the set of coverings $(U_i \mid i \in I(V))$ of V with $(U_i \mid i \in I)$ running through $\mathfrak{U}$. From the triangulation theorem [DK$_1$, 2.2] (cf. also the introduction of Chap. III) we know that $\mathfrak{U}$ (resp. $\mathfrak{V}$) is cofinal in the set of all finite open semialgebraic coverings of M (resp. V). (Consider coverings by open stars, cf. [D, §5], [D$_2$, §5]). Since $H^*(M,\mathcal{F})$ and $H^*(V,\mathcal{F})$ coincide with the Cech cohomology groups ([D, §5], [CC, Prop. 5]), it suffices to show that
i) the Cech groups $H^q((U_i \mid i \in I),\mathcal{F})$ are 0 for $q > n$
ii) the natural map

$$H^n((U_i \mid i \in I),\mathcal{F}) \longrightarrow H^n((U_i \mid i \in I(V)),\mathcal{F})$$

is surjective for every covering $(U_i \mid i \in I) \in \mathfrak{U}$. By definition $H^*((U_i \mid i \in I),\mathcal{F})$ is the homology of the complex $C^\bullet((U_i \mid i \in I),\mathcal{F})$ of alternating Cech cochains. Recall that

$$C^m((U_i \mid i \in I),\mathcal{F}) = \prod_{\substack{i_0,\ldots,i_m \in I \\ \text{pairwise different}}} \Gamma(U_{i_0} \cap \ldots \cap U_{i_m},\mathcal{F}).$$

We see that $C^q((U_i \mid i \in I),\mathcal{F}) = 0$ for $q > n$. This implies i) and shows that $H^n((U_i \mid i \in I),\mathcal{F})\{$ resp. $H^n((U_i \mid i \in I(V)),\mathcal{F})\}$ is a quotient of $C^n((U_i \mid i \in I),\mathcal{F})\{$ resp. $C^n((U_i \mid i \subset I(V)),\mathcal{F})\}$. Since the natural map $C^n((U_i \mid i \in I),\mathcal{F}) \longrightarrow C^n((U_i \mid i \in I(V)),\mathcal{F})$ is obviously surjective we also obtain ii). q.e.d.

Lemma 9.2. Let X be a topological space and $0 \longrightarrow \mathcal{F} \longrightarrow \mathcal{J}^0 \longrightarrow \mathcal{J}^1 \longrightarrow \ldots \longrightarrow \mathcal{J}^n \longrightarrow 0$ be an exact sequence of sheaves on X such that $\mathcal{J}^k$ is Φ-acyclic for $0 \leq k \leq n-1$ (Φ a support family on X). Then

$$H^q_\Phi(X,\mathcal{J}^n) = H^{q+n}_\Phi(X,\mathcal{F}) \quad \text{for} \quad q > 0.$$

Proof. Let $\mathcal{Z}^k = \ker(\mathcal{J}^k \longrightarrow \mathcal{J}^{k+1})$. The long exact cohomology sequences of the short exact sequences $0 \longrightarrow \mathcal{Z}^k \longrightarrow \mathcal{J}^k \longrightarrow \mathcal{Z}^{k+1} \longrightarrow 0$ show that

$$H^q_\Phi(X,\mathcal{Z}^n) = H^{q+1}_\Phi(X,\mathcal{Z}^{n-1}) = \ldots = H^{q+n}_\Phi(X,\mathcal{Z}^0)$$

for $q > 0$.

Let M be a geometric space over R and Φ be a family of supports on M.

Definiton 1. $\dim \Phi := \sup(\dim A \mid A \in \Phi)$.

Proposition 9.3. Suppose Φ is paracompactifying and $\dim \Phi = N < \infty$. Let $0 \longrightarrow \mathcal{F} \longrightarrow \mathcal{J}^0 \longrightarrow \mathcal{J}^1 \longrightarrow \ldots \longrightarrow \mathcal{J}^n \longrightarrow 0$ be an exact sequence of sheaves on M such that $\mathcal{J}^k$ is Φ-soft for $0 \leq k \leq n-1$. Then $\mathcal{J}^n$ is also Φ-soft.

Proof. It suffices to prove that $\mathcal{J}^n \mid A$ is soft for every $A \in \Phi$ (Prop. 4.1). Replacing M by A and applying Prop. 4.6 we may assume that M is affine semialgebraic, $\Phi = cld$ and $\dim M \leq n$. Then $H^q(M, \mathcal{G}) = 0$ for $q > n$ and every sheaf $\mathcal{G}$ on M (Lemma 9.1). We conclude from Lemma 9.2 by use of Theorem 4.9 that

$$H^1(M, (\mathcal{J}^n)_U) = H^{n+1}(M, \mathcal{F}_U) = 0$$

for every $U \in \mathring{\gamma}(M)$. Hence $\mathcal{J}^n$ is soft (Prop. 4.17).

Corollary 9.4. Suppose Φ is paracompactifying and $\dim \Phi = n < \infty$. Then $H^q_\Phi(M, \mathcal{F}) = 0$ for $q > n$ and every sheaf $\mathcal{F}$ on M.

Proof. Let $0 \longrightarrow \mathcal{F} \longrightarrow \mathcal{J}^0 \longrightarrow \mathcal{J}^1 \longrightarrow \ldots$ be a Φ-soft resolution of $\mathcal{F}$. By Prop. 9.3 the kernel $\ker(\mathcal{J}^n \longrightarrow \mathcal{J}^{n+1})$ is Φ-soft. Hence we have a Φ-soft (and hence Φ-acyclic) resolution of $\mathcal{F}$ of length n. This shows that $H^q_\Phi(M, \mathcal{F}) = 0$ for $q > n$.

Lemma 9.5. Let M be an affine semialgebraic space over R and $A \in \bar{\gamma}(M)$. Then $H^q_A(M, \mathcal{F}) = 0$ for $q > \dim M$ and every sheaf $\mathcal{F}$ on M.

Proof. The statement follows from (9.1) by use of the exact sequence
$\ldots \longrightarrow H^k_A(M, \mathcal{F}) \longrightarrow H^k(M, \mathcal{F}) \longrightarrow H^k(M \setminus A, \mathcal{F}) \longrightarrow H^{k+1}_A(M, \mathcal{F}) \longrightarrow \ldots$.

Proposition 9.6. Suppose M is of type (L) (i.e. every connected component of M is Lindelöf) and $\dim M = n < \infty$. Let $0 \longrightarrow \mathcal{F} \longrightarrow \mathcal{J}^0 \longrightarrow \mathcal{J}^1 \longrightarrow \ldots \longrightarrow \mathcal{J}^n \longrightarrow 0$ be an exact sequence of sheaves on M such that $\mathcal{J}^k$ is sa-flabby for $0 \leq k < n$. Then $\mathcal{J}^n$ is also sa-flabby.

Proof. We may assume that M is affine semialgebraic (Prop. 4.7). For $0 \leq k < n$ $\mathcal{J}^k$ is Φ-acyclic for any support family on M (Theorem 4.10). We conclude from Lemma 9.2 and Lemma 9.5 that

$$H^1_A(M, \mathcal{J}^n) = H^{n+1}_A(M, \mathcal{F}) = 0$$

for every $A \in \bar{\gamma}(M)$. Hence $\mathcal{J}^n$ is sa-flabby (Prop. 4.15).

Corollary 9.7. Assume that M is of type (L) and $\dim M = n < \infty$. Let $\mathcal{F}$ be a sheaf and Φ be an arbitrary family of supports on M. Then $H^q_\Phi(M, \mathcal{F}) = 0$ for $q > n$.

Proof. Let $0 \longrightarrow \mathcal{F} \longrightarrow \mathcal{J}^0 \longrightarrow \mathcal{J}^1 \longrightarrow \ldots$ be a sa-flabby resolution of $\mathcal{F}$. By Prop. 9.6 the kernel $\ker(\mathcal{J}^n \longrightarrow \mathcal{J}^{n+1})$ is sa-flabby. Hence $\mathcal{F}$ has a sa-flabby (and hence Φ-acyclic) resolution of length n. Thus $H^q_\Phi(M, \mathcal{F}) = 0$ for $q > n$. q.e.d.

Now we consider a regular and taut space M over R. Then $\tilde{M}^{\max}$ is a locally compact topological space (cf. I, §3).
Recall that $\dim_\Lambda \tilde{M}^{\max}$ is defined to be the smallest natural number n (or ∞) such that $H^q_\Psi(\tilde{M}^{\max}, \mathcal{F}) = 0$ for $q > n$ for every sheaf $\mathcal{F}$ and every family Ψ of supports on $\tilde{M}^{\max}$ which is paracompactifying in the classical sense (cf. [B, II.15.7]).

Lemma 9.8. $\dim_\Lambda \tilde{M}^{\max} \leq \dim M$.

Proof. Let Ψ be a family of supports on $\tilde{M}^{\max}$ which is paracompactifying in the classical sense. Let Φ be the family of closed subsets of $\tilde{M}$ which are contained in the closure $\bar{A}$ of some $A \in \Psi$. Then Φ is a family of supports on $\tilde{M}$ and $\Phi^m = \Phi \cap \tilde{M}^{max} = \Psi$. By Prop. 1.4 and Prop. 1.6 Φ is locally semialgebraic and paracompactifying. Let $\mathcal{G}$ be a sheaf on $\tilde{M}^{\max}$ and $i : \tilde{M}^{\max} \longrightarrow \tilde{M}$ be the inclusion map. Then $H^q_\Psi(\tilde{M}^{\max}, \mathcal{G}) = H^q_\Phi(\tilde{M}, i_*\mathcal{G})$ by Cor. 5.3 and $H^q_\Phi(M, i_*\mathcal{G}) = 0$ for $q > \dim M$ by Cor. 9.4. q.e.d.

Lemma 9.9. Assume that M is regular and taut. Then $\tilde{M}^{\max}$ is clc^∞_Λ (cf. [B, II.16.1] for the definition).

Proof. Let $x \in \tilde{M}^{\max}$ and $\mathfrak{A} := \{A \in \bar{\gamma}(M) \mid \tilde{A}$ is a neighbourhood of x in $\tilde{M}\}$. Then $\{\tilde{A} \mid A \in \mathfrak{A}\}$ is a fundamental system of neighbourhoods of x in $\tilde{M}$ by Prop. I.3.15. The triangulation theorem ([DK$_1$, §2]) shows that the sets $\tilde{A}, A \in \mathfrak{A}$ and A contractible, also form a fundamental system of neighbourhoods of x. But if A is contractible, then $H^q(\tilde{A} \cap \tilde{M}^{\max}, \Lambda) = H^q(\tilde{A}^{\max}, \Lambda) = H^q(A, \Lambda)$ is 0 for $q > 0$ and isomorphic to Λ for $q = 0$.
q.e.d.

§10 - Hypercohomology

In Chapter III it will sometimes be rather convenient to interpret the Borel-Moore-homology as a hypercohomology. Therefore we recall here the definition and some well known facts. A general reference is e.g. [H, Chap. I].
Let X be a topological space (e.g. an abstract locally semialgebraic space). We consider (cochain) complexes $\mathcal{F}^\bullet = (\ldots \longrightarrow \mathcal{F}^n \longrightarrow \mathcal{F}^{n+1} \longrightarrow \ldots \mid n \in \mathbf{Z})$ of sheaves on X. Every sheaf $\mathcal{F}$ on X is also considered as a complex $\mathcal{F}^\bullet : \mathcal{F}^0 := \mathcal{F}$ and $\mathcal{F}^k := 0$ for $k \neq 0$.
The cohomology sheaf $\frac{\mathrm{Ker}\,(\mathcal{F}^n \to \mathcal{F}^{n+1})}{\mathrm{Im}\,(\mathcal{F}^{n-1} \to \mathcal{F}^n)}$ of a complex $\mathcal{F}^\bullet$ is denoted by $\mathcal{H}^n(\mathcal{F}^\bullet)$.
Quite generally, if $A^\bullet$ and $B^\bullet$ are cochain complexes in an abelian category, then a chain map $f : A^\bullet \longrightarrow B^\bullet$ is said to be a *quasiisomorphism* if f induces an isomorphism $f_* : H^*(A^\bullet) \xrightarrow{\sim} H^*(B^\bullet)$ in cohomology. Thus a chain map $f : \mathcal{F}^\bullet \longrightarrow \mathcal{G}^\bullet$ between complexes of sheaves is a quasiisomorphism if f induces isomorphisms $\mathcal{H}^n(\mathcal{F}^\bullet) \longrightarrow \mathcal{H}^n(\mathcal{G}^\bullet) (n \in \mathbf{Z})$.
Let $\mathcal{F}^\bullet$ be a complex of sheaves and Φ be a family of supports on X. A quasiisomorphism $\mathcal{F}^\bullet \longrightarrow \mathcal{J}^\bullet$ is called an injective (flabby, Φ-acyclic) resolution of $\mathcal{F}^\bullet$ if every sheaf $\mathcal{J}^k$ is injective (flabby, Φ-acyclic). We denote the set of complexes of sheaves on X by $K(X)$ and the set of complexes $\mathcal{F}^\bullet$ which are bounded below (i.e. $\mathcal{F}^k = 0$ for k smaller than some bound $n_0 \in \mathbf{Z}$) by $K^+(X)$. If $\mathcal{F}^\bullet \in K^+(X)$, then $\mathcal{F}^\bullet$ has an injective resolution $\mathcal{F}^\bullet \longrightarrow \mathcal{J}^\bullet$. We may obtain $\mathcal{J}^\bullet$ as follows. Let $0 \longrightarrow \mathcal{F}^k \longrightarrow \mathcal{G}^{k\cdot}$ be the canonical injective resolution of $\mathcal{F}^k$ ([B, II.3]). Then the associated total complex $\mathcal{J}^\bullet$ of the double complex $(\mathcal{G}^{kl})$ (cf. [CE, IV, §4]) is an injective resolution of $\mathcal{F}^\bullet$. Injective resolutions are uniquely determined up to (chain) homotopy.
Let $\mathcal{F}^\bullet \in K^+(X)$ and $\mathcal{F}^\bullet \longrightarrow \mathcal{J}^\bullet$ be an injective resolution.

Definition 1. $\mathsf{H}^q_\Phi(X, \mathcal{F}^\bullet) := H^q(\Gamma_\Phi(X, \mathcal{J}^\bullet))$ is called the q-th *hypercohomology* group of X with coefficients in $\mathcal{F}^\bullet$ and supports in Φ.

Example 10.1. If $\mathcal{F}$ is a single sheaf, then $\mathsf{H}^q_\Phi(X, \mathcal{F}) = H^q_\Phi(X, \mathcal{F})$.

Remark 10.2. The hypercohomology may be computed by arbitrary Φ-acyclic resolutions $\mathcal{F}^\bullet \longrightarrow \mathcal{G}^\bullet$, i.e. there is a canonical isomorphism

$$H^q(\Gamma_\Phi(X, \mathcal{G}^\bullet)) \xrightarrow{\sim} \mathsf{H}^q_\Phi(X, \mathcal{F}^\bullet).$$

Now assume that $\dim_\Phi X = n < \infty$, i.e. that $H^q_\Phi(X, \mathcal{F}) = 0$ for $q > n$ for every sheaf $\mathcal{F}$ on X. Then every $\mathcal{F}^\bullet \in K(X)$ (not necessarily bounded) has a Φ-acyclic resolution $\mathcal{F}^\bullet \longrightarrow \mathcal{G}^\bullet$ and we may define $\mathsf{H}^q_\Phi(X, -)$ on whole $K(X)$ by setting $\mathsf{H}^q_\Phi(X, \mathcal{F}^\bullet) := H^q(\Gamma_\Phi(X, \mathcal{G}^\bullet))$. This definition does not depend on the choice of the Φ-acyclic resolution (cf. [H, Chap. I]).
If $\pi : Y \longrightarrow X$ is a continuous map, then π obviously induces a homomorphism

$$\pi^* : \mathsf{H}^*_\Phi(X, \mathcal{F}^\bullet) \longrightarrow \mathsf{H}^*_{\pi^{-1}(\Phi)}(Y, \pi^*\mathcal{F}^\bullet).$$

Now assume that either

a) $\mathcal{F}^\bullet \in K^+(X)$

or

b) $\mathcal{F}^\bullet \in K(X)$ and $\dim_\Phi X = n < \infty$.

Then we may compute $\mathsf{H}^*_\Phi(X,\mathcal{F}^\bullet)$ as follows: Let
$0 \longrightarrow \mathcal{F}^q \longrightarrow \tilde{\mathcal{L}}^0(\mathcal{F}^q) \longrightarrow \tilde{\mathcal{L}}^1(\mathcal{F}^q) \longrightarrow \dots$ be the canonical flabby Godement resolution of $\mathcal{F}^q (q \in \mathbb{Z})$. In case a) we set $\mathcal{L}^p(\mathcal{F}^q) = \tilde{\mathcal{L}}^p(\mathcal{F}^q)$ and in case b) we set

$$\mathcal{L}^p(\mathcal{F}^q) := \begin{cases} \tilde{\mathcal{L}}^p(\mathcal{F}^q) & \text{if } p < n \\ \mathrm{Ker}\,(\tilde{\mathcal{L}}^n(\mathcal{F}^q) \longrightarrow \tilde{\mathcal{L}}^{n+1}(\mathcal{F}^q)) & \text{if } p = n \\ 0 & \text{if } p > n \end{cases}$$

All sheaves $\mathcal{L}^p(\mathcal{F}^q)$ are Φ-acyclic (cf. Lemma 9.2). Let $\mathcal{J}^\bullet$ be the associated total complex of the double complex $(\mathcal{L}^p(\mathcal{F}^q) \mid p, q \in \mathbb{Z})$. Then $\mathcal{F}^\bullet \longrightarrow \mathcal{J}^\bullet$ is a Φ-acyclic resolution of $\mathcal{F}^\bullet$ and hence $\mathsf{H}^q_\Phi(X,\mathcal{F}^\bullet) = H^q(\Gamma_\Phi(X,\mathcal{J}^\bullet))$.
The first spectral sequence of the double complex $(\Gamma_\Phi(X,\mathcal{L}^p(\mathcal{F}^q)))$ converges and yields a spectral sequence (10.3) $H^p_\Phi(X,\mathcal{H}^q(\mathcal{F}^\bullet)) \Longrightarrow \mathsf{H}^{p+q}_\Phi(X,\mathcal{F}^\bullet)$.

Finally we consider an example. Let M be a geometric locally semialgebraic space over R and $X = \tilde{M}$. Let $\mathcal{F}^\bullet$ be a complex of sheaves and Φ be a family of supports on M. Of course we set $\mathsf{H}^*_\Phi(M,\mathcal{F}^\bullet) = \mathsf{H}^*_{\tilde{\Phi}}(\tilde{M},\tilde{\mathcal{F}}^\bullet)$ (if it is defined).
We assume that at least one of the following three conditions is satisfied:
a) $\mathcal{F}^\bullet$ is bounded below.
b) M is of type (L) and $\dim M < \infty$.
c) Φ is paracompactifying and $\dim \Phi < \infty$.
Note that $\dim_\Phi M < \infty$ in the cases b) and c) by Cor. 9.7 and Cor. 9.4. Hence the hypercohomology $\mathsf{H}^q_\Phi(M,\mathcal{F}^\bullet)$ is defined in all three cases.
Let $S \supset R$ be a real closed field extension. We use the notation of §6. (For example $\pi : \widetilde{M(S)} \longrightarrow \tilde{M}$ is the natural projection).
Consider the spectral sequences

$$H^p_\Phi(M,\mathcal{H}^q(\mathcal{F}^\bullet)) \Longrightarrow \mathsf{H}^{p+q}_\Phi(M,\mathcal{F}^\bullet)$$

$$H^p_{\Phi(S)}(M(S),\mathcal{H}^q(\mathcal{F}^\bullet(S))) \Longrightarrow \mathsf{H}^{p+q}_{\Phi(s)}(M(S),\mathcal{F}^\bullet(S)).$$

The projection π induces a homomorphism π^* from the first spectral sequence to the second. It is an isomorphism on the initial terms by Theorem 6.1. Hence it is also an isomorphism on the limit terms and we obtain

Theorem 10.4. The canonical map

$$\pi^* : \mathsf{H}^*_\Phi(M,\mathcal{F}^\bullet) \longrightarrow \mathsf{H}^*_{\Phi(S)}(M(S),\mathcal{F}^\bullet(S))$$

is an isomorphism.

In the same way we obtain by Theorem 6.10

Theorem 10.5. Assume that M is locally complete and $\dim M < \infty$. Let $\mathcal{F}^\bullet$ be a complex of sheaves on M.
a) The canonical homomorphism

$$\pi^* : \mathsf{H}^*_c(M,\mathcal{F}^\bullet) \longrightarrow \mathsf{H}^*_c(M(S),\mathcal{F}^\bullet(S))$$

is an isomorphism.
b) If M is paracompact, then

$$\pi^* : \mathrm{H}^*_{pc}(M, \mathcal{F}^\bullet) \longrightarrow \mathrm{H}^*_{pc}(M(S), \mathcal{F}^\bullet(S))$$

is also an isomorphism.

Recall that, in general, we have $c(M)(S) \subsetneq c(M(S))$ and $pc(M)(S) \subsetneq pc(M(S))$.

CHAPTER III: Semialgebraic Borel-Moore-homology

The spaces considered in this chapter are locally semialgebraic spaces over a real closed field R. Implicitly we assume that all occuring spaces are locally complete. Notice that locally complete spaces are regular and that regular semialgebraic spaces are affine (cf. [DK_3, I, §7], [Ro]).

We develop a homology theory for locally complete spaces over R with arbitrary supports and coefficients in an arbitrary sheaf. We fix a *principal ideal domain* Λ. All sheaves are assumed to be sheaves of Λ-modules and all tensor products are taken over Λ. (In some examples we take $\Lambda = \mathbf{Z}$ as the reader will easily realize). If M is a space over the field $\mathbf{R}$ of real numbers, then M_{top} denotes the set M equipped with its strong topology. Note that M_{top} is a topological subspace of $\tilde{M}$ (cf. I.5.4.d).
Since triangulations play a fundamental role in this chapter, we recall the *triangulation theorem*.

Theorem ([DK_3, II.4.4]). Let M be a regular and paracompact space and $(A_\lambda \mid \lambda \in I)$ be a locally finite family of locally semialgebraic subsets of M. Then there exists a strictly locally finite simplicial complex X (cf. [DK_3, I, §2]) and a locally semialgebraic isomorphism

$$\varphi : X \xrightarrow{\sim} M$$

such that $\varphi^{-1}(A_\lambda)$ is a subcomplex of X for every $\lambda \in I$. { Such an isomorphism is called a simultaneous triangulation of M and $(A_\lambda \mid \lambda \in I)$}.
If X is a simplicial complex, then $\bar{X}$ always denotes the closure of X (cf. [DK_3, I, §2]). The set of open simplices of X is denoted by $\Sigma(X)$. The set of connected components of a space M is denoted by $\pi_0(M)$.

§1 - Simplicial complexes and simplicial approximations

We consider an abstract simplicial complex $K = (E(K), S(K))$. The basic notions concerning abstract simplicial complexes may be found in [DK_3, II, §3]. $E(K)$ is the set of vertices and $S(K)$ is the set of simplices of K. By definition $S(K)$ consists of non-empty finite subsets s of $E(K)$ and we have $E(K) = \cup(s \mid s \in S(K))$. In contrast to classical combinatorial topology ([Sp, 3.1]), we do not assume that K is closed, i.e. if $s \in S(K)$ and $t \subset s$ is a face of s, then t not necessarily belongs to $S(K)$. The closure of K ([DK_3, II, §3]) is denoted by $\bar{K}$.

Definition 1. K is called *locally closed* if the following holds: Given simplices $s, t \in S(K)$ and a subset u of $E(K)$ with $s \subset u \subset t$, then $u \in S(K)$. (Note that u is always a simplex of the closure $\bar{K}$ of K).

Remark 1.1. The following statements are equivalent.

i) K is locally closed.
ii) K is open in its closure $\bar{K}$.
iii) $\partial K := \bar{K} \setminus K$ is a closed subcomplex of $\bar{K}$.
If K is locally finite then i) - iii) are also equivalent to
iv) The realization $\mid K \mid_R$ of K over R is a locally complete (locally semialgebraic) space over R.

i) $\Longleftrightarrow$ ii) $\Longleftrightarrow$ iii) follows from the definition, iii) $\Longleftrightarrow$ iv) from [DK_3, I, 7.3].

From now on we implicitly equip all occcuring abstract complexes K with a *partial order*. This means that the set $E(K)$ of vertices is partially ordered in such a way that every simplex $s \in S(K)$ is a totally ordered set. If L is a subcomplex of K, then $E(L)$ is assumed to be endowed with the partial order which is induced by the partial order of $E(K)$.

The set of simplices of K of dimension n is denoted by $S_n(K)$. Now we consider a strictly locally finite and locally closed abstract complex K. („Strictly locally finite" means that the closure $\bar{K}$ of K is locally finite, cf. [DK$_3$, II, §3, Def. 2 b]). If $s \in S_n(K), s = \{e_0, \ldots, e_n\}, e_i \in E(K)$ and $e_0 < e_1 < \ldots < e_n$, then we denote the ordered tupel $(e_0, \ldots, e_n)$ by $< s >$ or $< e_0, \ldots, e_n >$. Let $C_n(K, \Lambda)$ be the Λ-module of (possibly infinite) sums $\sum_{s \in S_n(K)} m_s \cdot < s >, m_s \in \Lambda$, if $n \geq 0$ and $C_n(K, \Lambda) = 0$ if $n < 0$. We define a boundary map

$$\partial : C_n(K, \Lambda) \longrightarrow C_{n-1}(K, \Lambda)$$

by

$$\partial < e_0, \ldots, e_n > := \sum_{i \in I} (-1)^i < e_0, \ldots, \hat{e}_i, \ldots, e_n >$$

where $I := \{i \in \mathbf{Z} \mid 0 \leq i \leq n, \{e_0, \ldots, \hat{e}_i, \ldots, e_n\} \in S_{n-1}(K)\}$
and

$$\partial (\sum_{S_n(K)} m_s \cdot < s >) := \sum_{S_n(K)} m_s \cdot \partial < s > .$$

Since K is locally finite, every $t \in S_{n-1}(K)$ is a face of at most finitely many $s \in S_n(K)$. Hence ∂ is a well defined map. (N.B. ∂ is just the classical boundary map, but we omit all faces which do not belong to K).

Lemma 1.2. $\partial \circ \partial = 0$.
We only have to verify that $\partial \circ \partial < s > = 0$. This is an easy exercise. But notice that the assumption „K locally closed" is essential at this point. Consider for instance the complex $\tilde{K}$ consisting of one 2-simplex $s = \{e_0, e_1, e_2\}$. one 1-simplex $\{e_0, e_1\}$ and one 0-simplex $t = \{e_0\}$. Suppose $e_0 < e_1 < e_2$. Then $\partial \circ \partial < s > = < t > \neq 0$. The complex $\tilde{K}$ is not locally closed since $\{e_0, e_2\}$ is not a simplex of $\tilde{K}$.

By Lemma 1.2 $\{(C_n, (K, \Lambda) \mid n \in \mathbf{Z}), \partial\}$ is a chain complex $C_\bullet(K, \Lambda)$. It is called the *oriented chain complex of K with closed supports* (and coefficients in Λ).
If K is a finite complex, then $C_\bullet(K, \Lambda) = C_\bullet(K, \mathbf{Z}) \otimes_{\mathbf{Z}} \Lambda$. This is not true in general. For example, $C_\bullet(K, \mathbf{Q}) \neq C_\bullet(K, \mathbf{Z}) \otimes_{\mathbf{Z}} \mathbf{Q}$ for an infinite complex K.

Lemma 1.3. Let L be a closed subcomplex of K.[1] Then $C_\bullet(L, \Lambda)$ is a subcomplex of $C_\bullet(K, \Lambda)$.
This is trivial.

If $L \subset K$ is a closed subcomplex, then $C_\bullet(K, L; \Lambda)$ is defined to be the quotient $C_\bullet(K, \Lambda)/C_\bullet(L, \Lambda)$. It is the same as $C_\bullet(K \setminus L, \Lambda)$.

[1] „Closed subcomplex" means: L is „closed in" K, i.e. $L = \bar{L} \cap K$.

In particular we have the identity

$$C_\bullet(K,\Lambda) = C_\bullet(\bar{K}, \partial K; \Lambda).$$

Now let K and K_1 be strictly locally finite locally closed abstract simplicial complexes and L and L_1 be closed subcomplexes of K and K_1. Let $\alpha : \bar{K} \to \bar{K}_1$ be a simplicial map which maps ∂K to ∂K_1 and $\bar{L}$ to $\bar{L}_1$. Notice that $\partial L = \partial K \cap \bar{L}$ and hence we have $\alpha(\partial L) \subset \partial K_1 \cap \bar{L}_1 = \partial L_1$.

We assume that α is proper ([DK$_3$, II, §5, Def. 2, 2a]). This means that every simplex $t \in S(\bar{K}_1)$ has a most finitely many preimages $s \in S(\bar{K})$ under α (or, equivalently, that $\alpha^{-1}(t)$ is a finite complex for every $t \in S(\bar{K}_1)$). Then α induces a map (from now on we often omit the coefficients Λ in our notation)

$$\tilde{\alpha}_\bullet : C_\bullet(\bar{K}) \to C_\bullet(\bar{K}_1)$$

with the following properties:

a) $\tilde{\alpha}_\bullet(C_\bullet(\partial K)) \subset C_\bullet(\partial K_1)$

b) $\tilde{\alpha}_\bullet(C_\bullet(\bar{L})) \subset C_\bullet(\bar{L}_1)$

c) $\tilde{\alpha}_\bullet(C_\bullet(\partial L)) \subset C_\bullet(\partial L_1)$

($\tilde{\alpha}$ is even a chain map, see Lemma 1.4 below).

It may be described as follows:
If $s \in S_n(\bar{K}), s = \{e_0, \ldots, e_n\}, e_0 < e_1 < \ldots < e_n$, then

$$\tilde{\alpha}_\bullet(< s >) := \begin{cases} 0 & \text{if } \dim \alpha(s) < n \\ \operatorname{sign} \pi \cdot < \alpha(s) > & \text{if } \alpha(s) \in S_n(\bar{K}_1) \end{cases}$$

where $\pi \in \gamma_n$ is the permutation with $\alpha(e_{\pi(0)}) \leq \alpha(e_{\pi(1)}) \leq \ldots \leq \alpha(e_{\pi(n)})$.
In general, $\tilde{\alpha}_\bullet(\sum\limits_{S_n(\bar{K})} n_s \cdot < s >) := \sum n_s \cdot \tilde{\alpha}_\bullet(< s >)$.
Since α is assumed to be proper, the sum and hence the map $\tilde{\alpha}_\bullet$ are well defined. Note that $C_\bullet(K, L) = C_\bullet(\bar{K}) / C_\bullet(\partial K) + C_\bullet(\bar{L})$.
Therefore $\tilde{\alpha}_\bullet$ induces a map

$$\alpha_\bullet : C_\bullet(K, L) \to C_\bullet(K_1, L_1).$$

Lemma 1.4. $\alpha_\bullet$ is a chain map.

Proof. We may assume that K and K_1 are finite complexes. Then $\bar{K}$ and $\bar{K}_1$ are finite closed complexes and $C_\bullet(\bar{K})$ and $C_\bullet(\bar{K}_1)$ are the classical oriented chain complexes of $\bar{K}$ and $\bar{K}_1$. In this situation the result is well known (cf. [Sp, 4.3]).

For any closed pair (K, L) of abstract locally finite and locally closed simplicial complexes we define simplicial homology groups. („Closed pair" means that L is a subcomplex of K which is closed in K).

Definition 2. $H_k(K, L; \Lambda) := H_k(C_\bullet(K, L; \Lambda))$ is called the k-th *homology group* of (K, L) *with coefficients in* Λ (and *closed supports*).

Obviously the proper semialgebraic map $\alpha : \bar{K} \to \bar{K}_1$ considered above induces a homomorphism

$$\alpha_* : H_*(K, L; \Lambda) \to H_*(K_1, L_1; \Lambda).$$

Now let X be a strictly locally finite and locally complete geometric simplicial complex over R (cf. [DK$_3$, I, §2]). Let $Y \subset X$ be a closed subcomplex.[2] The set of open simplices (n-simplices) of X is denoted by $\Sigma(X)(\Sigma_n(X))$. Let $K(X)$ and $K(Y)$ be the abstractions of X and Y (cf. [DK$_3$, II, §3]). We define

$$C_\bullet(X, Y; \Lambda) := C_\bullet(K(X), K(Y); \Lambda).$$

If $\sigma \in \Sigma_n(X)$ and $s \in S_n(K(X))$ is the abstraction of σ (i.e. s is the simplex of $K(X)$ whose realization $| s |_R$ is σ), then we often write $< \sigma >$ instead of $< s >$. So, for example, $C_n(X, \Lambda)$ is the module of (possibly infinite) sums $\sum\limits_{\sigma \in \Sigma_n(X)} n_\sigma \cdot < \sigma >, n_\sigma \in \Lambda$. The „boundary" $\bar{X} \setminus X$ of X is denoted by ∂X.

Let (X_1, Y_1) be another closed pair of strictly locally finite and locally complete geometric complexes over R. („Closed pair" means that Y is a closed subcomplex of X). Let $f : \bar{X} \to \bar{X}_1$ be a proper simplicial map (cf. [DK$_3$, I, 6.14]) such that $f(\partial X) \subset \partial X_1$ and $f(\bar{Y}) \subset \bar{Y}_1$. Then f is the realization $| \alpha |_R$ of a proper simplicial map $\alpha : \overline{K(X)} \to \overline{K(X_1)}$. This map α induces a chain map

$$\alpha_\bullet : C_\bullet(X, Y; \Lambda) \to C_\bullet(X_1, Y_1; \Lambda)$$

and a homomorphism

$$\alpha_* : H_*(X, Y; \Lambda) \to H_*(X_1, Y_1; \Lambda)$$

as we explained before. Here we used the notation $H_*(X, Y; \Lambda) := H_*(K(X), K(Y); \Lambda)$. The maps $\alpha_\bullet$ and α_* are also denoted by $f_\bullet$ and f_*.

Let M be a paracompact (and locally complete) geometric space over R and A be a closed locally semialgebraic subset of M. If $\varphi : X \xrightarrow{\sim} M$ is a simultaneous triangulation of M and A (cf. the introduction of this chapter) and $Y = \varphi^{-1}(A)$, then we use the following notation: $K(M, \varphi) := K(X), K(A, \varphi) := K(Y), C_\bullet(M, A, \varphi; \Lambda) := C_\bullet(X, Y; \Lambda), C_\bullet(A, \varphi; \Lambda) := C_\bullet(Y, \Lambda)$. The cycles { boundaries } of $C_\bullet(M, A, \varphi; \Lambda)$ are denoted by $Z_\bullet(M, A, \varphi; \Lambda) \{B_\bullet(M, A, \varphi; \Lambda)\}$.

In the following all geometric spaces over R are assumed to be paracompact (and locally complete, as always).

We consider a (locally semialgebraic) proper map $f : (M, A) \to (N, B)$ between closed pairs of spaces M and N (i.e. $A \in \bar{\mathcal{T}}(M), B \in \bar{\mathcal{T}}(N), f : M \to N$ proper, $f(A) \subset B$). We know that f is even semialgebraic ([DK$_3$, I.5.10]).

Definition 3. A *proper simplicial approximation to* f is a tripel (φ, ψ, g) consisting of simultaneous triangulations $\varphi : X \xrightarrow{\sim} M$ of M and A and $\psi : X_1 \xrightarrow{\sim} N$ of N and B and a simplicial map $g : \bar{X} \to \bar{X}_1$ with the following properties:

---

[2] „Closed subcomplex" means: Y is closed in X, i.e. $Y = \bar{Y} \cap X$.

i) The map $\psi^{-1} \circ f \circ \varphi : X \to X_1$ can be extended to a locally semialgebraic map $\bar{f} : \bar{X} \to \bar{X}_1$.

ii) g is a simplicial approximation to the map (of systems of complexes) $\bar{f} : (\bar{X}, \partial X, \bar{Y}) \to (\bar{X}_1, \partial X_1, \bar{Y}_1)$ where $Y := \varphi^{-1}(A)$ and $Y_1 := \psi^{-1}(B)$. (cf. [DK$_3$, III, §2, Def. 4]). { N.B. Since f is proper we have $\bar{f}^{-1}(X_1) = X$ and hence $\bar{f}(\partial X) \subset \bar{f}(\partial X_1)$}.

Recall that condition ii) means: g is a simplicial map $\bar{X} \to \bar{X}_1$ with $g(\partial X) \subset \partial X_1, g(\bar{Y}) \subset \bar{Y}_1$ and $\bar{f}(\mathrm{St}_{\bar{X}}(e)) \subset \mathrm{St}_{\bar{X}_1}(g(e))$ for every vertex e of $\bar{X}$.
The „completion" $\bar{f} : \bar{X} \to \bar{X}_1$ of f is a semialgebraic map. This implies that $g : \bar{X} \to \bar{X}_1$ is a proper simplicial map.

Let (φ, ψ, g) be a proper simplicial approximation to $f : (M, A) \to (N, B)$ and let (ψ, χ, k) be a proper simplicial approximation to another proper map $h : (N, B) \to (L, C)$. Then $(\varphi, \chi, k \circ g)$ is a proper simplicial approximation to $h \circ f : (M, A) \to (L, C)$.

Definition 4. Let $\psi : X_1 \xrightarrow{\sim} M$ and $\varphi : X \xrightarrow{\sim} M$ be simultaneous triangulations of M and A ($A \in \bar{\mathcal{T}}(M)$). We say that φ *refines* ψ (and write $\psi < \varphi$) if $\psi^{-1} \circ \varphi : X \to X_1$ maps every open simplex $\sigma \in \Sigma(X)$ of X into an open simplex τ of X_1. We say that φ *strictly refines* ψ (and write $\psi << \varphi$) if $\psi < \varphi$ and there is a proper simplicial approximation (φ, ψ, g) to $\mathrm{id}_M : M \to M$. (It immediately follows from [DK$_3$, III, 2.2] that (φ, ψ, g) is also a proper approximation to $\mathrm{id}_{(M,A)} : (M, A) \xrightarrow{\sim} (M, A)$).

Example 1.5. Let $\varphi' : X' \xrightarrow{\sim} M$ be the (first) barycentric subdivision of a simultaneous triangulation $\varphi : X \xrightarrow{\sim} M$ of M and A. Then $\varphi << \varphi'$.
In fact, let $g : \bar{X}' \longrightarrow \bar{X}$ be one of the simplicial maps which map the barycentre $\hat{\sigma}$ of a simplex $\sigma \in \Sigma(\bar{X})$ to some vertex of σ. Then (φ', φ, g) is a proper simplicial approximation to id_M.

Proper simplicial approximations always exist.

Proposition 1.6. Let $\psi : X_1 \xrightarrow{\sim} N$ be a simultaneous triangulation of N and B. Then there exists a proper simplicial approximation (φ, ψ, g) to $f : (M, A) \longrightarrow (N, B)$.

In view of Example 1.5 Proposition 1.6 follows from [DK$_3$, III.2.5] and the triangulation theorem.

Proposition 1.7. Let $\varphi : X \xrightarrow{\sim} M$ and $\psi : X_1 \xrightarrow{\sim} M$ be simultaneous triangulations of M and A. Then there exists a simultaneous triangulation $\chi : Y \xrightarrow{\sim} M$ of M and A with $\varphi << \chi$ and $\psi << \chi$.

Proof. We may assume that φ and ψ are barycentric subdivisions of other triangulations of (M, A) (Example 1.5). By [DK$_3$, II.5.1] there is a completion $i : M \hookrightarrow \bar{M}$ (which we regard as an inclusion map) and a map $h = (h_1, h_2) : \bar{M} \to \bar{X} \times \bar{Y}$ such that the diagram

$$\begin{array}{ccc} M & \xrightarrow{(\varphi^{-1}, \psi^{-1})} & X \times Y \\ i \downarrow & & \downarrow \\ \bar{M} & \xrightarrow{h} & \bar{X} \times \bar{Y} \end{array}$$

commutes. We choose a simultaneous triangulation $\tilde{\chi}: \bar{Y} \xrightarrow{\sim} \bar{M}$ of $\bar{M}, M, A$ and the locally finite families $(h_1^{-1}(\sigma) \mid \sigma \in \Sigma(\bar{X}))$ and $(h_2^{-1}(\tau) \mid \tau \in \Sigma(\bar{Y}))$. Let $Z := \tilde{\chi}^{-1}(M)$ and $\chi := \tilde{\chi} \mid Z \xrightarrow{\sim} M$. Then $\chi >> \varphi$ and $\chi >> \psi$ by [DK$_3$, III.2.5].

Now we consider a proper simplicial approximation (φ, ψ, g) to the given proper map $f:(M, A) \to (N, B)$ of closed pairs ($\varphi: X \xrightarrow{\sim} M, \psi: Y \xrightarrow{\sim} N$ triangulations). Then g induces a chain map $g_\bullet: C_\bullet(M, A, \varphi; \Lambda) \to C_\bullet(N, B, \psi; \Lambda)$. If (φ, ψ, k) is another approximation to f with respect to φ and ψ, then g and k are strictly contiguous ([DK$_3$, III.2.6]). This fact implies

Lemma 1.8. $g_\bullet$ and $k_\bullet$ are homotopic chain maps.

Proof. Let $\tilde{C}_\bullet(M, A, \varphi)$ and $\tilde{C}_\bullet(N, B, \psi)$ be the subcomplexes of $C_\bullet(M, A, \varphi)$ and $C_\bullet(N, B, \psi)$ consisting of the finite sums $\Sigma n_s \cdot < s >$ (all but finitely many n_s are 0). Then the maps $\tilde{g}_\bullet, \tilde{k}_\bullet : \tilde{C}_\bullet(M, A, \varphi) \rightrightarrows \tilde{C}_\bullet(N, B, \psi)$ obtained from $g_\bullet$ and $k_\bullet$ by restriction are homotopic. More precisely, since g and k are strictly contiguous, $g(\sigma)$ and $k(\sigma)$ span a closed simplex $\tau(\sigma)$ of $\bar{Y}$ for every $\sigma \in \Sigma(\bar{X})$. Moreover, $\overline{\tau(\sigma)} \subset \partial Y$ if $\sigma \subset \partial X$ and $\overline{\tau(\sigma)} \subset \overline{\psi^{-1}(B)}$ if $\sigma \subset \varphi^{-1}(A)$. Then there is a homotopy $\tilde{D}$ from $\tilde{g}_\bullet$ to $\tilde{k}_\bullet$ such that $\tilde{D}(< \sigma >)$ lies in the image of $C_\bullet(\tau(\sigma))$ under the natural projection $C_\bullet(\bar{Y}) \to C_\bullet(N, B, \psi)$ ([G, I.3.7.3]).
Clearly $\tilde{D}$ extends to a homotopy D from $g_\bullet$ to $k_\bullet$. We simply set

$$D(\Sigma n_\sigma \cdot < \sigma >) := \Sigma n_\sigma \cdot \tilde{D}(< \sigma >).$$

Then D is well-defined and $g_\bullet - k_\bullet = \partial D + D\partial$ since $\tilde{g}_\bullet - \tilde{k}_\bullet = \partial \tilde{D} + \tilde{D}\partial$.

In particular $g_\bullet$ and $k_\bullet$ induce the same homomorphism in the homology groups and we define

$$f_* := g_* = k_* : H_*(C_\bullet(M, A, \varphi; \Lambda)) \to H_*(C_\bullet(N, B, \psi; \Lambda)).$$

§2 - Semialgebraic Borel-Moore-homology of paracompact spaces

In this section all locally semialgebraic spaces are assumed to be paracompact (and locally complete). Let (M,A) be a closed pair of spaces over R (i.e. $A \in \bar{T}(M)$). The set of simultaneous triangulations of M and A is denoted by $\mathrm{Tr}(M,A)$. If $\varphi, \psi \in \mathrm{Tr}(M,A), \varphi : X \xrightarrow{\sim} M, \psi : X_1 \xrightarrow{\sim} M$ and $\varphi << \psi$ then we may choose a proper simplicial approximation (ψ, φ, g) of $\mathrm{id}_{(M,A)}$ and thus get an induced homomorphism $g_* : H_*(C_\bullet(M,A,\psi;\Lambda)) \to H_*(C_\bullet(M,A,\varphi;\Lambda))$. It does not depend on the choice of the approximation (Lemma 1.8). Therefore we may denote it by $\alpha_{\varphi,\psi}$. It turns out that $\alpha_{\varphi,\psi}$ is an isomorphism. We can even prove

Proposition 2.1. The chain map

$$g_\bullet : C_\bullet(M,A,\psi;\Lambda) \to C_\bullet(M,A,\varphi;\Lambda)$$

is a homotopy equivalence.

The proof of Prop. 2.1 requires some preparation. First we consider an arbitrary affine semialgebraic space M and a semialgebraic subset A of M. In [D] and [DK$_1$] we introduced homology groups $H_q(M,A;\Lambda)$. Since this homology turns out to be homology with complete supports (cf. Example 7.7) we denote these groups from now on by $H_q^c(M,A;\Lambda)$.

Lemma 2.2. Let $f : M \to N$ be a semialgebraic map between complete semialgebraic spaces M,N. Let $B \in \bar{\gamma}(N)$ and $A = f^{-1}(B)$. Assume that f induces an isomorphism $M \setminus A \to N \setminus B$. Then the induced map

$$H_*^c(M,A;\Lambda) \to H_*^c(N,B;\Lambda)$$

is an isomorphism.

Lemma 2.2 can be proved in the same way as the corresponding result for singular homology in classical topology (cf. [Wh, II.2.3]). In fact, mapping cylinders of proper semialgebraic maps and quotients M/A exist in the semialgebraic category provided A is complete ([DK$_3$, II, §10]). Every closed pair (M,A) of semialgebraic spaces has the homotopy extension property ([DK$_2$, §5], [DK$_3$, III, §1]). Moreover we have homotopy invariance and the excision property in our homology theory H_*^c (cf. [D], [DK$_1$]). Thus all tools are available which are necessary to copy the proof of the „classical Lemma 2.2".

Proof of Prop. 2.1 if M is semialgebraic:
The complexes X and X_1 are finite. Hence $C_\bullet(M,A,\psi)$ and $C_\bullet(M,A,\varphi)$ are free Λ-modules (we omit the coefficients Λ in the following) and it suffices to prove that $g_\bullet$ induces an isomorphism in homology ([Do, II.4.3]). Let Y be the closed subcomplex $\partial X \cup \varphi^{-1}(A)$ of $\bar{X}$ and Y_1 be the closed subcomplex $\partial X_1 \cup \psi^{-1}(A)$ of $\bar{X}_1$. Then $C_\bullet(M,A,\varphi) = C_\bullet(\bar{X})/C_\bullet(Y)$ and $C_\bullet(M,A,\psi) = C_\bullet(\bar{X}_1)/C_\bullet(Y_1)$. From the simplicial description of H_*^c in [DK$_1$, §3] we know that $H_*^c(\bar{X}_1, Y_1) = H_*(C_\bullet(\bar{X}_1)/C_\bullet(Y_1)) = H_*(C_\bullet(M,A,\psi))$ and $H_*^c(\bar{X},Y) = H_*(C_\bullet(\bar{X})/C_\bullet(Y)) = H_*(C_\bullet(M,A,\varphi))$. Let $f : \bar{X}_1 \to \bar{X}$ be the extension of $\varphi^{-1} \circ \psi : X_1 \to X$. Then f is proper, $f^{-1}(Y) = Y_1$ and f induces an isomorphism $\bar{X}_1 \setminus Y_1 \to \bar{X} \setminus Y$. Hence $f_* : H_*^c(\bar{X}_1, Y_1) \to H_*^c(\bar{X}, Y)$ is an isomorphism by Lemma 2.2. But f_* is defined by means of simplicial approximation, i.e., f_* is induced by $g_\bullet : C_\bullet(M,A,\psi) \to C_\bullet(M,A,\varphi)$ (cf. [DK$_1$, §3]). We see that $g_\bullet$ is indeed a homotopy equivalence.

For the general case we need two lemmas.

Lemma 2.3. Let B be a locally closed subset of M such that $Y := \overline{\varphi^{-1}(B)}$ is a subcomplex X. Let $Y_1 := \psi^{-1}(B)$. Then $g(\bar{Y}_1) \subset \bar{Y}, g(\partial Y_1) \subset \partial Y$ and $g(\overline{Y_1 \cap \psi^{-1}(A)}) \subset \overline{Y \cap \varphi^{-1}(A)}$. Hence $(\psi \mid Y_1, \varphi \mid Y, g \mid \bar{Y}_1)$ is a proper simplicial approximation to $\mathrm{id}_{(B, B\cap A)}$.

Proof. We consider a simplex $\sigma \in \Sigma(\bar{Y}_1)$ and choose some $\sigma_1 \in \Sigma(Y_1)$ with $\sigma \subset \bar{\sigma}_1$. Let $\tau \in \Sigma(Y)$ be the simplex containing $\varphi^{-1} \circ \psi(\sigma_1)$. The extension $\bar{X}_1 \to \bar{X}$ of $\varphi^{-1} \circ \psi$ is denoted by f. We have $f(\sigma) \subset f(\bar{\sigma}) \subset f(\bar{\sigma}_1) \subset \bar{\tau}$. Let $e_0, \dots, e_n$ be the vertices of σ. Since $f(e_k) \in \bar{\tau}$ and $f(e_k) \subset \mathrm{St}_{\bar{X}}(g(e_k))$, we see that $g(e_k) \in \bar{\tau} (0 \leq l \leq n)$. Hence $g(\sigma)$ is a face of τ and therefore it is contained in $\bar{Y}$.
Now we consider $Y_2 := \partial Y_1 \cap X_1$. Since B is locally closed, Y_2 is a closed subcomplex of X_1. From the first part of our proof we conclude that $g(\bar{Y}_2) \subset \overline{\varphi^{-1} \circ \psi(Y_2)} \subset \partial Y$. Now $\partial Y_1 = Y_2 \cup (\bar{Y}_1 \cap \partial X_1)$ and $g(\bar{Y}_1 \cap \partial X_1) \subset \bar{Y} \cap \partial X$ is also contained in ∂Y.
The last inclusion also follows from the first part of the proof (replace B by $B \cap A$).

Lemma 2.4. Let $C = \{(C_n \mid n \in \mathbb{Z}), \partial\}$ be a free chain complex of Λ-modules (i.e. every C_n is a free Λ-module; $\partial : C_n \to C_{n-1}$). Let C' and C'' be free subcomplexes of C such that $C' \cap C'' = 0$ and $C/(C' \oplus C'')$ is also free. Suppose $H_*(C) = H_*(C') = H_*(C'') = 0$. Let $D' : C' \to C'$ be a chain homotopy from $\mathrm{id}_{C'}$ to 0 (i.e. $\mathrm{id}_{C'} = \partial \circ D' + D' \circ \partial$) with $D' \circ D' = 0$. Then there is a chain homotopy $D : C \to C$ from id_C to 0 (i.e. $\mathrm{id}_C = \partial \circ D + D \circ \partial$) such that $D \circ D = 0, D \mid C' = D'$ and $D(C'') \subset C''$.

Proof. We denote the cycles of C by ZC and the boundaries by BC. By hypothesis we have $ZC = BC, ZC' = BC', ZC'' = BC''$. Let $s' : BC' \to C'$ be the restriction of D'. Since $\partial D' \partial c = \partial c - D' \partial \partial c = \partial c, s'$ is a section of $\partial : C' \to BC'$. The map D' is determined by s'. Namely, we have $C' = ZC' \oplus s'BC' = BC' \oplus s'BC'$, and if $c = \partial c_1 + s'(\partial c_2) \in C' (c_1, c_2, \in C')$, then $D'c = D' \partial c_1 + D' \circ D'(\partial c_2) = D' \partial c_1 = s'(\partial c_1)$ (N.B. $D' \circ D' = 0$).
Now we try to extend s' to a section $s : BC \to C$ of $C \xrightarrow{\partial} BC$ in such a way that $s(BC'') \subset C''$. Since $BC/(BC' \oplus BC'') = ZC/(ZC' \oplus ZC'') = ZC/ZC \cap (C' \oplus C'')$ is a subcomplex of the free complex $C/(C' \oplus C'')$, it is also a free chain complex and we conclude that $BC' \oplus BC''$ is a direct summand of BC. Hence $BC = BC' \oplus BC'' \oplus D$. Since all submodules of BC are free, we clearly can define an extension s as desired. Now we obtain B in the following way: We have $C = ZC \oplus sBC = BC \oplus sBC$. If $c = \partial c_1 + s\partial c_2 \in C$ then we define $D(c) := s\partial c_1$. We compute $\partial Dc + D\partial c = \partial s \partial c_1 + D(\partial s \partial c_2) = \partial c_1 + D\partial c_2 = \partial c_1 + s\partial c_2 = c$. Hence D is a homotopy from id_C to 0. Since $s \mid BC' = s'$, we have $D \mid C' = D'$. Finally we have by construction $D(C'') \subset C''$ and $D \circ D = 0$. The proof of Lemma 2.4 is finished.

Proof of Prop. 2.1 in the general case:
We may assume that M is connected. Then we choose finite closed subcomplexes $X_n, n \in \mathbb{N}$, of X such that $X = \cup(X_n \mid n \in \mathbb{N})$ and $X_n \cap X_m = \emptyset$ for $\mid m - n \mid > 1$ (cf. [DK$_3$, I.4.19]). Let $M_n := \varphi(X_n)$ and $A_n := M_n \cap A$. Then $C_\bullet(M_n, A_n, \varphi)$ and $C_\bullet(M_n, A_n, \psi)$ are subcomplexes of $C_\bullet(M, A, \varphi)$ and $C_\bullet(M, A, \psi)$. By Lemma 2.3 the chain map $g_\bullet$ induces chain maps

$$(g_n)_\bullet : C_\bullet(M_n, A_n, \psi) \to C_\bullet(M_n, A_n, \varphi)$$

and

$$(h_n)_\bullet : C_\bullet(M_n \cap M_{n-1}, A_n \cap A_{n-1}, \psi) \to C_\bullet(M_n \cap M_{n-1}, A_n \cap A_{n-1}, \varphi).$$

Let Con (g), Con (g_n) and Con $(h_n)(n \in \mathbb{N})$ be the mapping cones of $g_\bullet, (g_n)_\bullet$ and $(h_n)_\bullet$. ([Do, II.1]). It suffices to prove that Con (g) is homotopic to 0 ([Do, II.3.7]). Note that Con (g_n) and Con (h_n) are subcomplexes of Con (g). We already know from the semialgebraic case that $(g_n)_\bullet$ and $(h_n)_\bullet$ are homotopy equivalences for every $n \in \mathbb{N}$. Hence Con (g_n) and Con (h_n) are acyclic ([Do, II.2.14]).
Since Con (g_n) and Con $(g_n) \cap$ Con (g_{n-1}) = Con (h_n) are free and acyclic complexes and Con $(g_n)/($Con $(h_n) \oplus$ Con $(h_{n+1}))$ is free and Con $(g_{n-1}) \cap$ Con $(g_{n+1}) = 0$, we can apply Lemma 2.4. Successively we find chain homotopies

$$D_n : \text{Con}\,(g_n) \to \text{Con}\,(g_n)$$

from id $_{\text{Con}\,(g_n)}$ to $0 (n = 1, 2, \ldots)$ such that $D_n \circ D_n = 0, D_n(\text{Con}\,(g_n) \cap \text{Con}\,(g_{n+1})) \subset$ Con $(g_n) \cap$ Con (g_{n+1}) and $D_n \mid \text{Con}\,(g_n) \cap \text{Con}\,(g_{n-1}) = D_{n-1} \mid \text{Con}\,(g_n) \cap \text{Con}\,(g_{n-1})$.

Now we find the desired chain homotopy from id $_{\text{Con}\,(g)}$ to 0 as follows:
Recall that Con $(g)_r = C_{r-1}(M, A, \psi) \oplus C_r(M, A, \varphi)$. Hence an element $c \in$ Con $(g)_r$ has a (unique) representation

$$c = \sum_{s \in S_{r-1}(M \setminus A, \psi)} n_s \cdot < s > + \sum_{t \in S_r(M \setminus A, \varphi)} m_t \cdot < t > (n_s, m_t \in \Lambda).$$

Here we used the abbreviation $S_k(M \setminus A, \chi) := S_k(K(M \setminus A, \chi)) (\chi \in \{\varphi, \psi\})$. For every $s \in S_{r-1}(M \setminus A, \psi)$ (resp. $t \in S_r(M \setminus A, \varphi)$) we choose some $n \in \mathbb{N}$ with $s \in S(M_n, \psi)$ (resp. $t \in S(M_n, \varphi)$) and set

$$D(< s >) := D_n(< s >) \in \text{Con}\,(g_n) \subset \text{Con}\,(g)$$

(resp. $D(< t >) := D_n(< t >) \in$ Con $(g_n) \subset$ Con (g)). Notice that $D(< s >)$ (resp. $D(< t >)$ does not depend on the choice of n. Then we define

$$D(c) := \Sigma n_s \cdot D(< s >) + \Sigma m_t \cdot D(< t >).$$

D is well defined and id $_{\text{Con}\,(g)} = \partial D + D\partial$ since id $_{\text{Con}\,(g_n)} = \partial D_n + D_n \partial$ for every $n \in \mathbb{N}$. The proof of Proposition 2.1 is finished.

Now we define the Borel-Moore-homology of (M, A).

Definition 1. $H_k(M, A; \Lambda) := \varprojlim_{\varphi} H_k(C_\bullet(M, A, \varphi; \Lambda))$ is called the k-th (*semialgebraic Borel-Moore-) homology* group of (M, A) with closed supports and coefficients in Λ. Here φ runs through the set Tr (M, A) of simultaneous triangulations of M and A. For $\varphi << \psi$ the transition map is the isomorphism $\alpha_{\varphi,\psi} : H_k(C_\bullet(M, A, \psi)) \to H_k(C_\bullet(M, A, \varphi))$.[3]

Since all transition maps are isomorphisms we have

Proposition 2.5. The natural map

---

[3] We write $H_k(M, \Lambda)$ instead of $H_k(M, \emptyset; \Lambda)$.

$$H_*(M, A; \Lambda) \to H_*(C_\bullet(M, A, \varphi; \Lambda))$$

is an isomorphism for every $\varphi \in \mathrm{Tr}\,(M, A)$.

Hence we may compute the homology of (M, A) by means of any triangulation.

Remark 2.6. If M is semialgebraic and $\varphi: X \xrightarrow{\sim} M$ is a triangulation, then $C_n(M, \varphi; \Lambda)$ is free and is generated by the open (oriented) n-simplices $< \sigma >, \sigma \in \Sigma_n(X)$, of X. The boundary $\partial < \sigma >$ of $< \sigma >$ consists of those $(n-1)$-faces τ of σ which belong to X. In general $C_n(M, \varphi; \Lambda)$ consists of (possibly infinite) sums of open n-simplices. So intuitively the semialgebraic Borel-Moore-homology of M is computed by means of open simplices of M.

Examples 2.7. a) Let M be an open n-simplex σ. Then $C_k(M) = 0$ for $k \neq n$ and $C_n(M) = \Lambda \cdot < \sigma >$. Hence

$$H_k(M, \Lambda) \cong \begin{cases} \Lambda & \text{if } k = n \\ 0 & \text{if } k \neq n \end{cases}$$

But note that $H_k^c(M, \Lambda) = 0$ for $k > 0$ since M is contractible (cf. [DK$_1$], [D]).
b) If M is a complete semialgebraic space and $A \in \bar{\gamma}(M)$, then

$$H_*(M, A; \Lambda) = H_*^c(M, A; \Lambda).$$

This is clear from the definitions.
c) Let $M = \sqcup(M_i \mid i \in I)$ be the direct sum of a family of spaces M_i (cf. [DK$_3$, I.2.4]). Then we have

$$H_k(M, \Lambda) = \prod_{i \in I} H_k(M_i, \Lambda).$$

d) Since $C_k(M, \varphi) = 0$ for $k > \dim M$ we have $H_k(M, \Lambda) = 0$ for $k > \dim M$.

Now we consider an *open* locally semialgebraic subset U of some space M. Let $A := M \setminus U$ and choose a triangulation $\varphi \in \mathrm{Tr}\,(M, A)$. We have the exact sequence

$$0 \to C_\bullet(A, \varphi) \to C_\bullet(M, \varphi) \xrightarrow{r} C_\bullet(U, \varphi) \to 0 \qquad (*)$$

of chain complexes. Here r is the canonical projection $C_\bullet(M, \varphi) \to C_\bullet(M, \varphi)/C_\bullet(A, \varphi) = C_\bullet(U, \varphi)$. Applying r means „forgetting the simplices outside of U". It induces a restriction map

$$\mathrm{res} : H_*(M, \Lambda) \to H_*(U, \Lambda).$$

From (*) we obtain the long exact sequence

(2.8) $\ldots \to H_k(A, \Lambda) \to H_k(M, \Lambda) \xrightarrow{\mathrm{res}} H_k(U, \Lambda) \xrightarrow{\partial} H_{k-1}(A, \Lambda) \to \ldots.$

Observe that the connecting homomorphism
$\partial: H_k(U) = H_k(C_\bullet(U, \varphi)) \to H_{k-1}(C_\bullet(A, \varphi)) = H_{k-1}(A)$ is induced by the boundary map ∂ of $C_\bullet(M, \varphi)$. Of course (2.8) may be also written in the form

(2.8') $\ldots \to H_k(A, \Lambda) \to H_k(M, \Lambda) \to H_k(M, A; \Lambda) \xrightarrow{\partial} H_{k-1}(A, \Lambda) \to \ldots.$

Our homology theory is functorial with respect to proper maps. Let $f:(M,A)\to(N,B)$ be a proper map between closed pairs of spaces. We choose a proper simplicial approximation (φ,ψ,g) to f (Prop. 1.6) and define f_* by the commutative diagram

$$\begin{array}{ccc} H_*(M,A;\Lambda) & \xrightarrow{f_*} & H_*(N,B;\Lambda) \\ \cong\downarrow & & \downarrow\cong \\ H_*(C_\bullet(M,A,\varphi;\Lambda)) & \xrightarrow{g_*} & H_*(C_\bullet(N,B,\psi;\Lambda)). \end{array}$$

It is easily seen by the results of §1 that f_* does not depend on the choice of the approximation. If $g:(N,B)\to(L,C)$ is another proper map, then $(g\circ f)_*=g_*\circ f_*$. The exact sequence (2.8') is functorial with respect to proper maps $f:(MA)\to(N,B)$.

Example 2.9. Let $(M_\alpha \mid \alpha\in I)$ be a locally finite family of closed locally semialgebraic subsets of M. Let $i_\alpha:M_\alpha\hookrightarrow M$ be the inclusion map and $i:\sqcup(M_\alpha\mid\alpha\in I)\to M$ be the map whose restriction to M_α is i_α for every $\alpha\in I$. ($\sqcup$ denotes the disjoint union). Then i is a proper semialgebraic map and hence induces a homomorphism

$$i_*:\prod_{\alpha\in I} H_*(M_\alpha,\Lambda)\to H_*(M,\Lambda).$$

If $(z_\alpha)_{\alpha\in I}\in\prod\limits_{\alpha\in I} H_k(M_\alpha,\Lambda)$, then the image $i_*((z_\alpha)_{\alpha\in I})$ is denoted by $\sum\limits_{\alpha\in I}(i_\alpha)_*z_\alpha$ or simply by $\sum\limits_{\alpha\in I} z_\alpha$. (N.B. If I is finite, then $\sum\limits_{\alpha\in I}(i_\alpha)_*z_\alpha$ is the usual sum of the homology classes $(i_\alpha)_*z_\alpha$).

Properly homotopic maps induce the same homomorphism in homology.

Definition 2. Two maps $f,g:(M,A)\rightrightarrows(N,B)$ are said to be *properly homotopic* if there is a proper semialgebraic map (a „homotopy") $H:(M\times[0,1],A\times[0,1])\to(N,B)$ such that $H_0=f$ and $H_1=g$. Here $H_t\,(t\in[0,1])$ denotes the map $x\mapsto H(x,t)$.[4] N.B. This implies that f and g are proper.

Theorem 2.10. Let $f,g:(M,A)\rightrightarrows(N,B)$ be properly homotopic maps. Then

$$f_*=g_*:H_*(M,A;\Lambda)\to H_*(N,B;\Lambda).$$

Proof. It suffices to consider the maps $f=i_0, g=i_1$ where $i_t:(M,A)\to(M\times[0,1],A\times[0,1])$ is the map $x\mapsto(x,t)$. Let $\varphi\in\mathrm{Tr}(M,A),\varphi:X\xrightarrow{\sim}M$. Identifying M with X by φ we may assume that M is a simplicial complex. We have a natural triangulation ψ of $M\times[0,1]$, the „prism decomposition" (cf. [Go, I.3.7], [ES, II.8]). With respect to the triangulations φ and ψ the maps i_0 and i_1 are simplicial. It is well known that the induced maps $(i_0)_\bullet,(i_1)_\bullet:C_\bullet(M,A,\varphi)\rightrightarrows C_\bullet(M\times[0,1],A\times[0,1],\psi)$ are homotopic ([Go, I.3.7]) Hence $i_{0*}=i_{1*}$.

We have Mayer-Vietoris sequences.

---

[4] Of course $[0,1]$ denotes the unit interval *in* R.

Proposition 2.11. Let M be the union of two closed locally semialgebraic subsets A and B. Then there is an exact sequence
$\dots \to H_n(A \cap B, \Lambda) \to H_n(A, \Lambda) \oplus H_n(B, \Lambda) \to H_n(M, \Lambda) \to H_{n-1}(A \cap B, \Lambda) \to \dots.$

Proof. We choose a simultaneous triangulation φ of M, A and B. Then $C_\bullet(M, \varphi) = C_\bullet(A, \varphi) + C_\bullet(B, \varphi)$ and $C_\bullet(A, \varphi) \cap C_\bullet(B, \varphi) = C_\bullet(A \cap B, \varphi)$. Now we obtain the exact sequence by [Sp, 4.6].

Theorem 2.12. Assume $R = \mathbf{R}$. Then the semialgebraic Borel-Moore-homology $H_*(M, A; \Lambda)$ coincides with the topological Borel-Moore-homology ${}^{BM}H_*(M_{\text{top}}, A_{\text{top}}, \Lambda)$ of the pair $(M_{\text{top}}, A_{\text{top}})$ as defined in [BM] (or [B, Chap. V]).

This will be proved later in a more general situation (§11). Here we sketch an elementary proof. Since $H_*(M, A) = H_*(M \setminus A)$ in both cases we may assume that $A = \emptyset$. Let ${}^{ls}H_*(M_{\text{top}})$ be the homology of M_{top} computed by locally finite singular chains ([Sw]), [B, I.7]). It is known that ${}^{ls}H_*(M_{\text{top}}) = {}^{BM}H_*(M_{\text{top}})$ ([B, V.ii.16]). Now there is a natural map $H_*(M) \to {}^{ls}H_*(M_{\text{top}})$. Let $M \subset \bar{M}$ be a completion of M ([DK$_3$, II. §2]). Using the long exact sequence (2.8), its analogue for singular homology and the Five Lemma we reduce to the case where $M = \bar{M}$. Let $\varphi: X \xrightarrow{\sim} M$ be a triangulation. Then $K(M, \varphi)$ is a closed (hence classical) abstract simplicial complex and it is known in this case that ${}^{ls}H_*(M_{\text{top}}) = H_*(C_\bullet(M, \varphi))$.

Again we consider a closed pair (M, A) of spaces over R. Let $S \supset R$ be a real closed overfield of R. By base extension we obtain a pair $(M(S), A(S))$ over S (cf. II, §6). Let $\varphi \in \mathrm{Tr}(M, A)$. Tarski's principle implies that the base extension φ_S of φ ([DK$_3$, I.2.10]) is a simultaneous triangulation of $M(S)$ and $A(S)$. Since $K(M, \varphi)$ (resp. $K(A, \varphi)$) is equal to $K(M(S), \varphi_S)$ (resp. $K(A(S), \varphi_S)$) we obtain

Theorem 2.13. There is a canonical iosomorphism

$$H_*(M, A; \Lambda) \cong H_*(M(S), A(S); \Lambda).$$

This result will also be generalized later (§10).

If $U \subset V$ are open locally semialgebraic subsets of a space M, then, as explained above, we have a restriction map $H_*(V, \Lambda) \xrightarrow{\text{res}} H_*(U, \Lambda)$. Hence $U \mapsto H_k(U, \Lambda)$ is a presheaf on M. The associated sheaf is denoted by $\mathcal{H}_k(M, \Lambda)$ or simply by $\mathcal{H}_k$.

Proposition 2.14. Assume $\dim M \leq n$. Then

$$\Gamma(U, \mathcal{H}_n) = H_n(U, \lambda)$$

for every $U \in \mathring{T}(M)$.

Proof. Let $(U_i \mid i \in I)$ be a locally finite open covering of U. We choose a simultaneous triangulation $\varphi: X \xrightarrow{\sim} M$ of M, U and the family $(U_i \mid i \in I)$ ([DK$_3$, II, §4]). Let $[z_i] \in H_n(U_i, \Lambda)$ be homology classes which are represented by cycles $z_i \in C_n(U_i, \varphi)$. Assume that $[z_i] \mid U_i \cap U_j = [z_j] \mid U_i \cap U_j$ for all pairs (i, j). Since there are no simplicies of dimension $> n$ this implies that $z_i \mid U_i \cap U_j = z_j \mid U_i \cap U_j$ for all pairs (i, j). We see that the cycles z_i glue together to form a cycle $z \in C_n(U, \varphi)$. Then $[z] \mid U_i = [z_i]$ for every $i \in I$. These considerations prove Prop. 2.14.

§3 - Fundamental classes of manifolds and real varieties

We consider a paracompact (and locally complete) space M over R. Let $\varphi: X \xrightarrow{\sim} M$ be a triangulation of M. By $\text{St}_\varphi(x)$ we denote the „star neighbourhood" $\varphi(\text{St}_X(\varphi^{-1}(x)))$ of x in M with respect to $\varphi\,(x \in M)$. Recall that $\text{St}_X(\varphi^{-1}(x))$ is the union of those open simplices of X which contain $\varphi^{-1}(x)$ in their closure. Let $\psi: Y \to M$ be another triangulation of M and assume that ψ refines φ. Then $\text{St}_\psi(x) \subset \text{St}_\varphi(x)$ for every $x \in M$.

Proposition 3.1. The restriction map

$$H_*(\text{St}_\varphi(x), \Lambda) \xrightarrow{\text{res}} H_*(\text{St}_\psi(x), \Lambda)$$

is an isomorphism.

Proof. Identifying M and X by φ we assume that $M = X$ and φ is the identity map. Let $U = \text{St}_X(x) = \text{St}_\varphi(x)$ and $V = \text{St}_\psi(x)$. Let Y_1 be the subcomplex $\psi^{-1}(U)$ of Y and $\psi_1: Y_1 \xrightarrow{\sim} U$ be the restriction of ψ.
First we study the special case where x is a vertex of X and ψ_1 extends to a triangulation $\bar\psi_1: \bar Y_1 \xrightarrow{\sim} \bar U$ of the closed complex $\bar U$ such that $\bar\psi_1$ refines the given (tautological) triangulation of $\bar U$. For every vertex e of $\bar Y_1$ we choose a vertex $g(e)$ of the open simplex σ of $\bar U$ which contains $\bar\psi_1(e)$. If $e \neq \psi_1^{-1}(x)$, then we choose $g(e)$ different from x. This implies that $g(e)$ is contained in $\bar U \setminus U$ if $e \neq \psi_1^{-1}(x)$. Note that $g(\psi_1^{-1}(x)) = x$.

The map $e \mapsto g(e)$ yields a simplicial map between pairs of simplicial complexes

$$g: (\bar Y_1, \bar Y_1 \setminus \psi^{-1}(V)) \to (\bar U, \bar U \setminus U)$$

(cf. [DK$_3$, III, proof of Thm. 2.5]). Let r be the map of pairs $g \circ \bar\psi_1^{-1}: (\bar U, \bar U \setminus V) \to (\bar U, \bar U \setminus U)$. By $i: (\bar U, \bar U \setminus U) \to (\bar U, \bar U \setminus V)$ we denote the map of pairs induced by $\text{id}_{\bar U}$. It is easy to see that the map $r \circ i: (\bar U, \bar U \setminus U) \to (\bar U, \bar U \setminus U)$ (resp. $i \circ r: \bar U \to \bar U$) is linearly homotopic to $\text{id}_{(\bar U, \bar U \setminus U)}$ (resp. $\text{id}_{\bar U}$). Let $H: \bar U \times [0,1] \to \bar U$ be the linear homotopy $H(x,t) := (1-t)\cdot x + t \cdot r(x)$ from $\text{id}_{\bar U}$ to $i \circ r$. Since $\bar V$ is isomorphic to the cone over $\bar V \setminus V$ with vertex x, we have an obvious retraction $R: \bar U \setminus \{x\} \to \bar U \setminus V$. We observe that $H((\bar U \setminus \{x\}) \times [0,1]) \subset \bar U \setminus \{x\}$ and that $R \circ (H \mid (\bar U \setminus V) \times [0,1]): (\bar U \setminus V) \times [0,1] \to \bar U \setminus V$ is a homotopy from $\text{id}_{\bar U \setminus V}$ to "$i \circ r$"$: \bar U \setminus V \to \bar U \setminus V$.
Now we conclude from Theorem 2.10 that $i_*: H_*(\bar U) \to H_*(\bar U)$ and $i_*: H_*(\bar U \setminus U) \to H_*(\bar U \setminus V)$ are isomorphisms. (The first map is just the identity). Hence

$$i_*: H_*(\bar U, \bar U \setminus U) \to H_*(\bar U, \bar U \setminus V)$$

is also an isomorphism. Now Prop. 3.1 follows (in the special case) since $H_*(\bar U, \bar U \setminus U) = H_*(U)$ and $H_*(\bar U, \bar U \setminus V) = H_*(V)$ and the diagram

$$\begin{array}{ccc} H_*(U) & \xrightarrow{\text{res}} & H_*(V) \\ \| & & \| \\ H_*(\bar U, \bar U \setminus U) & \xrightarrow[\sim]{i_*} & H_*(\bar U, \bar U \setminus V) \end{array} \qquad (*)$$

commutes (cf. §2).

Now we drop the assumption that ψ_1 extends to a triangulation $\bar{\psi}_1 : \bar{Y}_1 \to \bar{U}$ which refines the given triangulation of $\bar{U}$. We choose simultaneous triangulations $\tilde{\varphi}$ and χ of $(\bar{X}, X)$ and $(\bar{Y}, Y)$ such that
1) $\tilde{\varphi}$ refines the given triangulation of $\bar{X}$.
2) χ refines the given triangulation of $\bar{Y}$.
3) $\tilde{\psi} := \psi \circ (\chi \mid \chi^{-1}(Y)) : \chi^{-1}(Y) \xrightarrow{\sim} X$ refines $\tilde{\varphi} \mid \tilde{\varphi}^{-1}(X) : \tilde{\varphi}^{-1}(X) \xrightarrow{\sim} X$.
4) the closure $\overline{\mathrm{St}}_{\tilde{\varphi}}(x)$ of $\mathrm{St}_{\tilde{\varphi}}(x)$ is contained in X.
This is possible by the triangulation theorem. Then we consider the commutative diagram

$$\begin{array}{ccc} H_*(\mathrm{St}_{\varphi}(x)) & \xrightarrow{\text{res}} & H_*(\mathrm{St}_{\psi}(x)) \\ \text{res} \downarrow & & \downarrow \text{res} \\ H_*(\mathrm{St}_{\tilde{\varphi}}(x)) & \xrightarrow{\text{res}} & H_*(\mathrm{St}_{\tilde{\psi}}(x)) \end{array} \qquad (**)$$

From the special case studied above we know that the vertical arrows and the lower horizontal arrow are isomorphisms. Hence the upper horizontal arrow is also an isomorphism.
Finally, if x is not a vertex of X, then we may subdivide $\mathrm{St}_X(x)$ linearly and obtain a new triangulation $\tilde{\varphi} : X_1 \xrightarrow{\sim} X$ such that $\tilde{\varphi}^{-1}(x)$ is a vertex of X_1 and $\mathrm{St}_{\tilde{\varphi}}(x) = \mathrm{St}_{\varphi}(x)$. In the same way we choose a refinement $\tilde{\psi}_1 : Y_2 \xrightarrow{\sim} X$ of ψ such that $\tilde{\psi}_1^{-1}(x)$ is a vertex of Y_2 and $\mathrm{St}_{\tilde{\psi}_1}(x) = \mathrm{St}_{\psi}(x)$. The only problem now is that $\tilde{\psi}_1$ not necessarily refines $\tilde{\varphi}$. But we may choose a triangulation $\tilde{\psi}$ of X which refines $\tilde{\varphi}$ and $\tilde{\psi}_1$. Again we consider the commutative diagram (**). As before we already know that the vertical arrows and the lower horizontal arrow are isomorphisms. Hence $H_*(\mathrm{St}_{\varphi}(x)) \xrightarrow{\text{res}} H_*(\mathrm{St}_{\psi}(x))$ is also an isomorphism. q.e.d.

Definition 1. M is called a Λ-*homology-manifold* of dimension n if $\mathcal{H}_k(M, \Lambda)$ (cf. §2) is 0 for $k \neq n$ and $\mathcal{H}_n(M, \Lambda)$ is locally constant with stalks isomorphic to Λ. A $\mathbb{Z}$-homology-manifold is simply called a *homology-manifold*.

Remarks 3.2. a) Let $\varphi : X \xrightarrow{\sim} M$ be a triangulation. Then we easily derive from Prop. 3.1 that M is a Λ-homology-manifold of dimension n if and only if $H_k(\mathrm{St}_{\varphi}(x), \Lambda) = 0$ for $k \neq n$ and $H_n(\mathrm{St}_{\varphi}(x), \Lambda) \cong \Lambda$ for every $x \in M$.
b) Suppose M is a homology-manifold of dimension n. By the universal coefficient theorem ([Sp, 5.2.8]) we conclude that

$$H_k(\mathrm{St}_{\varphi}(x), \Lambda) = H_k(\mathrm{St}_{\varphi}(x), \mathbb{Z}) \otimes_{\mathbb{Z}} \Lambda$$

where φ is a triangulation of some open semialgebraic neighbourhood U of $x \in M$. Hence $\mathcal{H}_k(M, \Lambda) = \mathfrak{H}_k(M, \mathbb{Z}) \otimes_{\mathbb{Z}} \Lambda$ and we see that M is also a Λ-homology-manifold.
c) If M is a Λ-homology-manifold of dimension n, then M is pure of dimension n ([DK$_3$, I, §3, Def. 4]). To prove this let $x \in M$ and $\varphi : X \xrightarrow{\sim} U$ be a triangulation of an open semialgebraic neighbourhood U of x in M. Then $H_n(\mathrm{St}_{\varphi}(x)) \cong \Lambda$. Hence $\mathrm{St}_X(\varphi^{-1}(x))$ contains an n-simplex. On the other hand if $\dim \mathrm{St}_X(\varphi^{-1}(x))$ were $k > n$, then $H_k(\mathrm{St}_X(y), \Lambda) = H_k(\sigma, \Lambda)$ would be isomorphic to Λ for every open k-simplex σ of $\mathrm{St}_X(\varphi^{-1}(x))$ and every $y \in \sigma$. This is impossible.

Example 3.3. Let M be an n-dimensional manifold over R. (This means that every $x \in M$ has a neighbourhood $U \in \mathring{\gamma}(M)$ which is semialgebraically isomorphic to the open n-ball $B^n = \{x \in R^n \mid \ \| x \| < 1\}$). Then M is an n-dimensional homology-manifold.

Proof. We may assume that M is semialgebraic and choose a triangulation $\varphi : X \xrightarrow{\sim} M$. Let $x \in M$ and $U \in \mathring{\gamma}(M)$ be a neighbourhood of x such that $U \subset \mathrm{St}_{\varphi}(x)$ and the pair $(\bar{U}, U)$ is isomorphic to the pair $(\bar{B}, B)$ with B the open and $\bar{B}$ the closed n-ball. First triangulating $M \setminus U$ suitably and then adding the cone over $\bar{U} \setminus U$ with vertex x we obtain a triangulation ψ of M such that ψ refines φ and $U = \mathrm{St}_{\psi}(x)$. Now we conclude from Prop. 3.1 that

$$H_k(\mathrm{St}_{\varphi}(x), \mathbf{Z}) \cong H_k(U, \mathbf{Z}) \cong H_k(B^n, \mathbf{Z}) = \begin{cases} \mathbf{Z} & k = n \\ 0 & k \neq n \end{cases}.$$

By Remark 3.2.a we see that M is a homology-manifold of dimension n.

Let M be a Λ-homology-manifold of dimension n.

Definition 2. M is said to be Λ-*orientable* if $\mathfrak{H}_n(M, \Lambda)$ is isomorphic to the constant sheaf Λ_M.

Remarks 3.4. (In d) and e) we assume that M is even a $\mathbf{Z}$-homology-manifold).

a) Since $\dim M = n$, we have $\Gamma(U, \mathcal{H}_n) = H_n(U, \Lambda)$ for every $U \in \mathring{T}(M)$ (by Prop. 2.14).

b) Suppose M is connected. Then M is Λ-orientable if and only if $H_n(M, \Lambda) \cong \Lambda$.

c) If M is Λ-orientable, $U \in \mathring{T}(M)$ and both spaces M and U are connected, then the restriction map $H_n(M, \Lambda) \to H_n(U, \Lambda)$ is an isomorphism.

d) If M is $\mathbf{Z}$-orientable, then M is Λ-orientable.

e) M is always $\mathbf{Z}/_2$-orientable.

f) If M is simply connected, then M is Λ-orientable.

Proofs. b) The sheaf $\mathcal{H}_n(M, \Lambda)$ is locally isomorphic to Λ_M. It is constant if and only if $H_n(M, \Lambda) = \Gamma(M, \mathcal{H}_n) \cong \Lambda$.
c) is trivial.
d) is clear since $\mathcal{H}_n(M, \Lambda) = \mathcal{H}_n(M, \mathbf{Z}) \otimes_{\mathbf{Z}} \Lambda$.
e) We may assume that M is connected. Let $(U_i \mid i \in I)$ be an admissible covering of M by open connected sets $U_i \in \mathring{\gamma}(M)$ with $H_n(U_i, \mathbf{Z}/_2) \cong \mathbf{Z}/_2$. Let z_i be the non zero element of $H_n(U_i, \mathbf{Z}/_2)$. From c) we conclude that $z_i \mid U_i \cap U_j = z_j \mid U_i \cap U_j$. Hence the elements z_i glue together to form an element $z \in H_n(M, \mathbf{Z}/_2)$. We see that $H_n(M, \mathbf{Z}/_2) \cong \mathbf{Z}/_2$.
f) Every locally constant sheaf on a simply connected space is constant.

Definition 3. A Λ-*orientation* of M is an isomorphism $\chi : \Lambda_M \xrightarrow{\sim} \mathcal{H}_n$. Let $\bar{1}$ be the element $(\ldots, 1, \ldots) \in \prod_{\pi_0(M)} \Lambda = \Gamma(M, \Lambda_M)$. Then $[M] := \chi(\bar{1}) \in \Gamma(M, \mathfrak{H}_n) = H_n(M, \Lambda)$ is called the *fundamental class* of the Λ-*oriented* homology-manifold M. Not that $[M] \mid U$ is a generator of $H_n(U, \Lambda) \cong \Lambda$ for every open connected locally semialgebraic subset U of M.

If M is Λ-orientable and connected, then we obviously have a 1-1-correspondence between the Λ-orientations of M and the units of Λ. In particular every homology-manifold M has a unique $\mathbf{Z}/_2$-orientation and a unique fundamental class $[M] \in H_n(M, \mathbf{Z}/_2)$.
Now we derive a result which was proved for manifolds by an elementary but lengthy argument [DK, Thm. 13.2].

Proposition 3.5. Let M be a homology-manifold of dimension n and $A \in \bar{\mathcal{T}}(M)$ be a closed locally semialgebraic subset with dim $A \leq n-2$. Then the natural map $\pi_0(M \setminus A) \to \pi_0(M)$ is bijective.

Proof. Let $U := M \setminus A$. Since dim $A \leq n-2$ we have $H_n(A, \mathbb{Z}/_2) = H_{n-1}(A, \mathbb{Z}/_2) = 0$. We conclude from the long exact sequence 2.8 that $H_n(M, \mathbb{Z}/_2) \to H_n(U, \mathbb{Z}/_2)$ is an isomorphism. Since $H_n(M, \mathbb{Z}/_2) \cong \prod_{\pi_0(M)} \mathbb{Z}/_2$ and $H_n(U, \mathbb{Z}/_2) \cong \prod_{\pi_0(U)} \mathbb{Z}/_2$, our claim follows. q.e.d.

At the end of this section we generalize results of Borel and Haefliger on fundamental classes of real varieties over the field $\mathbb{R}$ of real numbers ([BH]) to the case where the ground field R is an arbitrary real closed field.
So let us consider a reduced separated scheme V over R which is locally of finite type. The set $V(R)$ of R-rational points of V is a locally complete locally semialgebraic space over R. We assume that $V(R)$ is Zariski-dense in V and that every irreducible component V' of V has dimension n. The Zariski-open subset of regular points of V is denoted by V_{reg}. The space $V_{\text{reg}}(R)$ is a locally semialgebraic manifold of dimension n (cf. [DK, §13]). If V is of finite type (i.e. quasicompact), then $V(R)$ is semialgebraic and hence even an affine semialgebraic space. Since we have not yet defined Borel-Moore-homology for non-paracompact locally complete spaces, we must assume here that $V(R)$ is a paracompact space. This is e.g. the case if V admits a covering $(U_i \mid i \in I)$ by Zariski-open affine subsets such that, for fixed i, $U_i \cap U_j = \emptyset$ for all but finitely many $j \in I$. But note that this assumption is not really necessary.

Definition 4. A homology class $z \in H_n(V(R), \mathbb{Z}/_2)$ is called a *fundamental class* of V (mod 2) if the restriction $z \mid V_{\text{reg}}(R)$ is the fundamental class $[V_{\text{reg}}(R)] \in H_n(V_{\text{reg}}(R), \mathbb{Z}/_2)$ of the manifold $V_{\text{reg}}(R)$.

Let $A := V(R) \setminus V_{\text{reg}}(R) = V_{\text{sing}}(R)$. Since dim $A \leq n-1$, we obtain from (2.8) the exact sequence

(3.6) $\quad 0 \to H_n(V(R), \mathbb{Z}/_2) \to H_n(V_{\text{reg}}(R), \mathbb{Z}/_2) \to H_{n-1}(A, \mathbb{Z}/_2)$.

We see that V posesses at most one fundamental class.

Theorem 3.7. V has a fundamental class $[V] \in H_n(V(R), \mathbb{Z}/_2)$.

Proof. Let $(V_k \mid k \in I)$ be the family of irreducible components of V. Then $V_{\text{reg}} = \sqcup(V_{k,\text{reg}} \mid k \in I)$ (disjoint union) and $H_*(V_{\text{reg}}(R), \mathbb{Z}/_2) = \prod_{k \in I} H_*(V_{k,\text{reg}}(R), \mathbb{Z}/_2)$. Let $i_k : V_k(R) \hookrightarrow V(R)$ be the inclusion map. Since every affine R-variety W has only finitely many irreducible components, the family $(V_k(R) \mid k \in I)$ is a locally finite family of closed locally semialgebraic subsets of $V(R)$.

Now suppose we have found a fundamental class $[V_k] \in H_n(V_k(R), \mathbb{Z}/_2)$ for every $k \in I$. Then it is obvious that the sum $\sum_{k \in I} (i_k)_*[V_k]$ (cf. Example 2.9) is a fundamental class of V. Hence we may assume from now on that V is irreducible. If V is normal, then the singular locus V_{sing} of V has at most dimension $n-2$. Thus dim $A \leq n-2$ and $H_{n-1}(A, \mathbb{Z}/_2) = 0$. In this case we conclude from the exact sequence 3.6 that the fundamental class $[V]$ exists.

In general we consider a normalization $\pi : V' \to V$ of V. Then $\pi_r = \pi \mid V'(R) : V'(R) \to V(R)$ is a proper semialgebraic map and hence induces a homomorphism

$$(\pi_R)_* : H_n(V'(R), \mathbb{Z}/_2) \to H_n(V(R), \mathbb{Z}/_2).$$

Let $[V'] \in H_n(V'(R), \mathbb{Z}/_2)$ be the fundamental class of V'. Since "π" $: \pi^{-1}(V_{\text{reg}}) \to V_{\text{reg}}$ is an isomorphism, we see that $(\pi_R)_*[V']$ is a fundamental class of V. q.e.d.

We saw in the proof that $H_n(V(R), \mathbb{Z}/_2) \xrightarrow{\sim} H_n(V_{\text{reg}}(R), \mathbb{Z}/_2)$ if V is irreducible and normal. This implies

Proposition 3.8. Let V be irreducible and normal and $(U_i \mid i \in I)$ be the family of connected components of $V_{\text{reg}}(R)$. Then $[V] \mid U_i$ generates $H_n(U_i, \mathbb{Z}/_2) \cong \mathbb{Z}/_2$ and the canonical homomorphism

$$H_n(V(R), \mathbb{Z}/_2) \to \prod_{i \in I} H_n(U_i, \mathbb{Z}/_2)$$

given by the restriction maps is an isomorphism.

Finally we mention another characterization of the fundamental class.

Proposition 3.9. Let $z \in H_n(V(R), \mathbb{Z}/_2)$. Then z is the fundamental class of V if and only if the image of z generates the stalk $\mathcal{H}_n(V(R), \mathbb{Z}/_2)_x$ for every $x \in V_{\text{reg}}(R)$.

The proof of Prop. 3.9 is trivial. If V is a complete variety over R, then these results were already obtained (by a completely different approach) in [DK$_1$, §8].

§4 - Subdivision and the sheaf of simplicial chains

Let M be a paracompact (and locally complete) space M over R and $\varphi : X \xrightarrow{\sim} M$ and $\psi : X_1 \xrightarrow{\sim} M$ be triangulations of M with $\varphi << \psi$. We proved in §2 that the induced map

$$g_\bullet : C_\bullet(M,\psi;\Lambda) \to C_\bullet(M,\varphi;\Lambda)$$

is a homotopy equivalence for every proper simplicial approximation (ψ,φ,g) to id_M. Now we are looking for a homotopy inverse of $g_\bullet$. Intuitively it is clear what we have to do. We have to „subdivide" every oriented n-simplex $<\sigma>$ of the triangulation φ, i.e. we have to replace $<\sigma>$ by the sum of those n-simplices τ of X_1 which are mapped into $\varphi(\sigma)$ by ψ. These simplices τ must be oriented in the right way.
Let $\sigma \in \Sigma_n(X)$ be an n-simplex. Then $\varphi(\sigma)$ is a locally closed subset of M and we conclude from Lemma 2.3 that g yields by restriction a proper simplicial approximation g_σ (with respect to φ,ψ) to $\mathrm{id}_{\varphi(\sigma)}$. Hence g induces a chain map $(g_\sigma)_\bullet : C_\bullet(\varphi(\sigma),\psi) \to C_\bullet(\varphi(\sigma),\varphi)$. (Again we omit the coefficients Λ in our notation). Observe that $C_k(\varphi(\sigma),\varphi) = 0$ if $k \neq n$ and $C_n(\varphi(\sigma),\varphi) = \Lambda \cdot <\sigma>$. The homomorphism $(g_\sigma)_* : H_*(C_\bullet(\varphi(\sigma),\psi)) \to H_*(C_\bullet(\varphi(\sigma),\varphi))$ is an isomorphism (Prop. 2.1). Since $\dim \varphi(\sigma) = n$, we have $H_n(C_\bullet(\varphi(\sigma),\psi)) = Z_n(\varphi(\sigma),\psi)$ and $H_n(C_\bullet(\varphi(\sigma),\varphi)) = Z_n(\varphi(\sigma),\varphi) = \Lambda \cdot <\sigma>$ $\{Z_\bullet(-,-)$ denotes the cycles of $C_\bullet(-,-)$, cf. §1 $\}$. Note that $C_\bullet(\varphi(\sigma),\psi)$ is a Λ-submodule (but no subcomplex) of $C_\bullet(M,\psi)$. Now we define

$$\mathrm{sd}\,(<\sigma>) := \mathrm{sd}_{\psi,\varphi}(<\sigma>) := (g_\sigma)_*^{-1} <\sigma> \in Z_n(\varphi(\sigma),\psi) \subset C_n(M,\psi).$$

If $c = \Sigma\ n_\sigma \cdot <\sigma> \in C_n(M,\varphi)$ is an arbitrary chain, then we set $\mathrm{sd}\,(c) := \mathrm{sd}_{\psi,\varphi}(c) := \Sigma n_\sigma \cdot \mathrm{sd}\,(<\sigma>) \in C_n(M,\psi)$.

By Lemma 1.8 the subdivision map $\mathrm{sd}_{\psi,\varphi}$ does not depend on the choice of the approximation g.

Definition 1. Let $c = \Sigma n_\sigma \cdot <\sigma> \in C_n(M,\varphi)$. The closure of $\cup(\varphi(\sigma) \mid n_\sigma \neq 0)$ is called the *support* of c. It will be denoted by $\mathrm{supp}\,(c)$ or simply by $|c|$.

Lemma 4.1. $\mathrm{sd} : C_\bullet(M,\varphi) \to C_\bullet(M,\psi)$ is a chain map.

Proof. It suffices to check that $\partial\,\mathrm{sd}\,(<\sigma>) = \mathrm{sd}\,(\partial <\sigma>)$ for some $\sigma \in \Sigma_n(X)$. Let $g_{\partial\sigma}$ be the proper simplicial approximation to $\mathrm{id}_{|\partial<\sigma>|}$ which is induced by g (Lemma 2.3). We have the commutative diagram

$$\begin{array}{ccc} Z_n(\varphi(\sigma),\varphi) & \xrightarrow{\partial} & Z_{n-1}(|\partial <\sigma>|,\varphi) \\ (g_\sigma)_* \uparrow\cong & & \cong\uparrow (g_{\partial\sigma})_* \\ Z_n(\varphi(\sigma),\psi) & \xrightarrow{\partial} & Z_{n-1}(|\partial <\sigma>|,\psi). \end{array}$$

The horizontal arrows are just the connecting map $H_n(\varphi(\sigma)) \xrightarrow{\partial} H_{n-1}(|\partial <\sigma>|)$ in the long exact sequence 2.8 of the closed pair $(\varphi(\sigma), |\partial <\sigma>|)$. Let λ be the inverse map of $(g_{\partial\sigma})_*$. Since ∂ is given by the boundary map of $C_\bullet(M,\varphi)$ (cf. §2), it suffices to show that λ coincides with the restriction of sd to $Z_{n-1}(|\partial <\sigma>|,\varphi)$. This follows from the commutative diagram

$$\begin{array}{ccc} Z_{n-1}(|\partial<\sigma>|,\varphi) & \stackrel{j}{\hookrightarrow} & \bigoplus_{i=1}^{r} Z_{n-1}(\varphi(\tau_i),\varphi) \\ \uparrow (g_{\partial\sigma})_* & & \uparrow \prod (g_{\tau_i})_* \\ Z_{n-1}(|\partial<\sigma>|,\psi) & \stackrel{j'}{\hookrightarrow} & \bigoplus_{i=1}^{r} Z_{n-1}(\varphi(\tau_i),\psi). \end{array}$$

Here $\tau_1,\ldots,\tau_r$ are the open $(n-1)$-faces of σ (which belong to X) and j (resp. j') is the product of the restriction maps res from $|\partial<\sigma>|$ to $\varphi(\tau_i)$ $(1\le i\le r)$.

Remark 4.2. Obviously we have $g_\bullet \circ \mathrm{sd} = \mathrm{id}$. Since $g_\bullet$ is a homotopy equivalence, we see that the subdivision map sd is a homotopy inverse of $g_\bullet$, i.e. $\mathrm{sd}\circ g_\bullet \simeq \mathrm{id}$.

Subdivision is transitive. Let $\chi : X_2 \xrightarrow{\sim} M$ be another triangulation and assume $\psi << \chi$.

Lemma 4.3. $\mathrm{sd}_{\chi,\varphi} = \mathrm{sd}_{\chi,\psi}\circ \mathrm{sd}_{\psi,\varphi}$.

Proof. It suffices to prove that $\mathrm{sd}_{\chi,\varphi}(<\sigma>) = \mathrm{sd}_{\chi,\psi}(\mathrm{sd}_{\psi,\varphi}(<\sigma>))$ for some $\sigma\in\Sigma_n(X)$. Let (χ,ψ,h) be a proper simplicial approximation to id_M. Then $(\chi,\varphi,g\circ h)$ is also a proper simplicial approximation to id_M. Our claim follows from the commutative diagram

$$\begin{array}{ccccc} Z_n(\varphi(\sigma),\varphi) & \stackrel{(g_\sigma)_*}{\longleftarrow} & Z_n(\varphi(\sigma),\psi) & \stackrel{j}{\hookrightarrow} & \bigoplus_{i=1}^{r} Z_n(\psi(\tau_i),\psi) \\ (g_\sigma\circ h_\sigma)_* \nwarrow & & (h_\sigma)_*\uparrow & & \uparrow \prod (h_{\tau_i})_* \\ & & Z_n(\varphi(\sigma),\chi) & \stackrel{j'}{\hookrightarrow} & \bigoplus_{i=1}^{r} Z_n(\psi(\tau_i),\chi). \end{array}$$

Here $\tau_1,\ldots,\tau_r$ are the n-simplices of X_1 which are contained in $\psi^{-1}\varphi(\sigma)$ and j (resp. j') is the product of the restriction maps from $\varphi(\sigma)$ to $\psi(\tau_i)(1\le i\le r)$.

If $\varphi,\psi,\in \mathrm{Tr}(M)$ and $\varphi<\psi$ (i.e. ψ is a (not necessarily strict) refinement of φ), then we define a subdivision map as follows. We choose some $\chi\in\mathrm{Tr}(M)$ with $\varphi<<\chi$ and $\psi<<\chi$ (Prop. 1.7) and a proper simplicial approximation (χ,ψ,g) to id_M. Then we set $\widetilde{\mathrm{sd}}_{\psi,\varphi} := g_\bullet\circ \mathrm{sd}_{\chi,\varphi} : C_\bullet(M,\varphi)\to C_\bullet(M,\psi)$.

Lemma 4.4. $\widetilde{\mathrm{sd}}_{\varphi,\psi}$ does not depend on the choice of χ and the choice of g.

Proof. Let σ be an n-simplex of X $(\varphi : X \xrightarrow{\sim} M)$. We choose a proper simplicial approximation (χ,φ,k) to id_M and consider the diagram

$$\begin{array}{ccc} Z_n(\varphi(\sigma),\varphi) & & Z_n(\varphi(\sigma),\psi) \\ (k_\sigma)_* \nwarrow\cong & & \cong\nearrow (g_\sigma)_* \\ & Z_n(\varphi(\sigma),\chi) & \end{array}$$

By definition $\widetilde{\mathrm{sd}}_{\psi,\varphi}(<\sigma>)$ is just $(g_\sigma)_*\circ(k_\sigma)_*^{-1}<\sigma>$ (considered as an element of $C_n(M,\psi)$). By Lemma 1.8 $(g_\sigma)_*$ and hence $\widetilde{\mathrm{sd}}_{\psi,\varphi}(<\sigma>)$ does not depend on the choice of

g.
Now let χ' be another triangulation with $\chi' >> \varphi, \chi' >> \psi$ and let (χ', ψ, g') be a proper simplicial approximation to id_M. We want to show that using $\text{sd}_{\chi',\varphi}$ and this approximation we obtain the same subdivision map $\widetilde{\text{sd}}_{\varphi,\psi}$ as before. From Remark 4.2 and Lemma 4.3 we conclude that we may assume that $\chi' >> \chi$. Then we take some proper simplicial approximation (χ', χ, k) to id_M. From above we know that we may replace g' by $g \circ k$. Now our claim follows since $\text{sd}_{\chi',\varphi} = \text{sd}_{\chi',\chi} \circ \text{sd}_{\chi,\varphi}$ and $k_\bullet \circ \text{sd}_{\chi',\chi} = \text{id}$ (Remark 4.2, Lemma 4.3).

In particular we see that $\widetilde{\text{sd}}_{\psi,\varphi}$ coincides with $\text{sd}_{\psi,\varphi}$ if $\psi >> \varphi$. Therefore we write from now on $\text{sd}_{\psi,\varphi}$ instead of $\widetilde{\text{sd}}_{\psi,\varphi}$. It is an easy exercise to verify the following analogue of Lemma 4.3.

Lemma 4.5. If $\varphi < \psi < \chi$, then

$$\text{sd}_{\chi,\varphi} = \text{sd}_{\chi,\psi} \circ \text{sd}_{\psi,\varphi}.$$

Working with fixed triangulations $\varphi < \psi$, we usually write sd instead of $\text{sd}_{\psi,\varphi}$.
Now let $\varphi : X \xrightarrow{\sim} M$ and $\psi : Y \xrightarrow{\sim} M$ be triangulations with $\varphi < \psi$.

Remark 4.6. If M is semialgebraic, then $C_\bullet(M, \varphi; \Lambda) = C_\bullet(M, \varphi; \mathbb{Z}) \otimes_{\mathbb{Z}} \Lambda$. Let $\text{sd}_{\psi,\varphi} : C_\bullet(M, \varphi; \mathbb{Z}) \to C_\bullet(M, \psi; \mathbb{Z})$ be the subdivision map defined over $\mathbb{Z}$. Then it is obvious that $\text{sd}_{\psi,\varphi} \otimes \text{id}_\Lambda : C_\bullet(M, \varphi; \Lambda) \to C_\bullet(M, \psi; \Lambda)$ is the subdivision map defined over Λ.

Subdivision preserves supports. More precisely we have

Lemma 4.7. Let $\sigma \in \Sigma_n(X)$ be an open n-simplex. Then

$$\text{sd}_{\psi,\varphi}(<\sigma>) = \sum_{\substack{\tau \in \Sigma_n(Y) \\ \psi(\tau) \subset \varphi(\sigma)}} \varepsilon_\tau < \tau >$$

with $\varepsilon_\tau \in \{\pm 1\}$ for each τ. In particular

$$\text{supp}(c) = \text{supp}(\text{sd}(c))$$

for every $c \in C_n(M, \varphi; \Lambda)$.

Proof. We may assume that $\Lambda = \mathbb{Z}$. Consider some $\tau \in \Sigma_n(Y)$ with $\psi(\tau) \subset \varphi(\sigma)$. Let $x \in \psi(\tau)$. We regard $<\sigma>$ as a generator of $Z_n(\varphi(\sigma), \varphi, \mathbb{Z}) = H_n(\varphi(\sigma), \mathbb{Z})$. Note that $\varphi(\sigma)$ and $\psi(\tau)$ are the star neighbourhoods of x in $\varphi(\sigma)$ with respect to the triangulations φ and ψ. By Prop. 3.1 the restriction $<\sigma> |\psi(\tau)$ is a generator of $H_n(\psi(\tau), \mathbb{Z}) = Z_n(\psi(\tau), \psi; \mathbb{Z}) = \mathbb{Z} \cdot <\tau>$ and hence $\text{sd}_{\psi,\varphi}(<\sigma>)|\psi(\tau) = \pm <\tau>$.

Corollary 4.8. sd is an injective map.

Remark 4.9. Let $\varphi : X \xrightarrow{\sim} M$ and $\psi : X_1 \xrightarrow{\sim} M$ be triangulations and assume that $\varphi < \psi$ and $\psi < \varphi$. This does not imply that the triangulations φ, ψ are equivalent (cf. [DK$_3$, II, §4]). But the subdivision map $\text{sd}_{\psi,\varphi}: C_\bullet(M,\varphi) \to C_\bullet(M,\psi)$ is an isomorphism with inverse map $\text{sd}_{\varphi,\psi}$. If $\sigma \in \Sigma(X), \tau \in \Sigma(X_1)$ and $\varphi(\sigma) = \psi(\tau)$, then $\text{sd}_{\psi,\varphi}(<\sigma>) = \pm <\tau>$. (The sign depends on the orientations).

Now we are able to define *sheaves of simplicial chains* on M. Let $U \in \mathring{\mathcal{T}}(M)$ and define

$$\Gamma(U, \Delta_k(M,\Lambda)) = \varinjlim C_k(U,\varphi;\Lambda) \quad (k = 0,1,2,\ldots).$$

Here the direct limit is taken over all triangulations φ of U with transition maps $\text{sd}_{\psi,\varphi}: C_\bullet(U,\varphi;\Lambda) \to C_\bullet(U,\psi;\Lambda)$ if $\psi > \varphi$. If $V \in \mathring{\mathcal{T}}(M)$ and $V \subset U$, then the set of simultaneous triangulations of U and V denoted by $\text{Tr}\,(U,V)$ is cofinal in the set $\text{Tr}\,(U)$ of triangulations of U (cf. the triangulation theorem [DK$_3$, II.4.4]). If $\varphi \in \text{Tr}\,(U,V)$, then we have the restriction map

$$C_\bullet(U,\varphi) \to C_\bullet(V,\varphi)$$

(forget the simplices outside of V, cf. §2). In this way we obtain a restriction map

$$\Gamma(U, \Delta_k(M,\Lambda)) \to \Gamma(V, \Delta_k(M,\Lambda)).$$

Hence $\Delta_k(M,\Lambda): U \mapsto \Gamma(U,\Delta_k(M,\Lambda))$ is a presheaf on M. We will often simply write Δ_k or $\Delta_k(M)$ instead of $\Delta_k(M,\Lambda)$. Δ_k is called the *sheaf of simplicial k-chains* on M (with coefficients in Λ). In fact it is a sheaf

Proposition 4.10. Δ_k is a sheaf.

Proof. Let $U \in \mathring{\mathcal{T}}(M)$ and $(U_i|i \in I)$ be a locally finite covering of U by open locally semialgebraic subsets. Let $[c_i] \in \Gamma(U_i,\Delta_k)$ $(i \in I)$ be sections represented by chains $c_i \in C_k(U_i,\varphi_i), \varphi_i \in \text{Tr}\,(U_i)$, such that $[c_i]|U_i \cap U_j = [c_j]|U_i \cap U_j$ for every pair $(i,j) \in I \times I$. By the triangulation theorem we find a triangulation $\varphi: X \xrightarrow{\sim} U$ of U such that $\varphi|\varphi^{-1}(U_i): \varphi^{-1}(U_i) \xrightarrow{\sim} U_i$ refines φ_i for every $i \in I$. Subdividing the chains c_i we may assume that $\varphi_i = \varphi|\varphi^{-1}(U_i)$ for every $i \in I$. Now it is clear by Cor. 4.8 that $c_i|U_i \cap U_j = c_j|U_i \cap U_j$ for all i,j. Hence the chains c_i glue together to form a chain $c \in C_k(U,\varphi)$. Obviously c yields an element $[c] \in \Gamma(U,\Delta_k)$ with $[c]|U_i = [c_i]$ for every i, and $[c]$ is the only element of $\Gamma(U,\Delta_k)$ which restricts to $[c_i]$ for every $i \in I$.

Lemma 4.11. Let $U \in \mathring{\mathcal{T}}(M)$. Then

$$\Delta_k(U,\Lambda) = \Delta_k(M,\Lambda)|U.$$

This observation is trivial. Nevertheless it allows us to extend the definition of Δ_k to arbitrary spaces M.

From now on let M be a locally complete but not necessarily paracompact space over R. By Lemma 4.11 there is a unique sheaf $\mathcal{F}$ of simplicial chains on M such that $\mathcal{F}|U = \Delta_k(U,\Lambda)$ for every open semialgebraic subset U of M. If M is paracompact, then $\mathcal{F}$ is the sheaf $\Delta_k(M,\Lambda)$. Therefore we also denote this sheaf $\mathcal{F}$ by $\Delta_k(M,\Lambda)$ or simply by $\Delta_k(M)$ or Δ_k $(k = 0,1,2,\ldots)$. The sheaves of simplicial chains have a remarkable property which immediately follows from the definition and the triangulation theorem.

Proposition 4.12. Let $A \in \bar{\mathcal{T}}(M)$. Then

$$\Gamma_A(M, \Delta_k(M,\Lambda)) = \Gamma(A, \Delta_k(A,\Lambda)).$$

Remark 4.13. Let $U \in \mathring{\mathcal{T}}(M)$. Then we have a canonical section

$$\mathrm{cl} = \mathrm{cl}_{M,U} : \Gamma(U, \Delta_k) \to \Gamma(M, \Delta_k)$$

of the restriction map $\Gamma(M, \Delta_k) \to \Gamma(U, \Delta_k)$.
Namely, let $(M_\lambda | \lambda \in I)$ be an admissible covering of M by open semialgebraic subsets. From the triangulation theorem we conclude that

$$\Gamma(U \cap M_\lambda, \Delta_k) = \varinjlim_{\mathrm{Tr}(M_\lambda, U \cap M_\lambda)} C_k(U \cap M_\lambda, \varphi)$$

and

$$\Gamma(M_\lambda, \Delta_k) = \varinjlim_{\mathrm{Tr}(M_\lambda, U \cap M_\lambda)} C_k(M_\lambda, \varphi).$$

Since $C_\bullet(U \cap M_\lambda, \varphi)$ is a submodule of $C_\bullet(M_\lambda, \varphi)$, we have a canonical monomorphism

$$\mathrm{cl}_\lambda : \Gamma(U \cap M_\lambda, \Delta_\bullet) \to \Gamma(M_\lambda, \Delta_\bullet).$$

It is easily seen that these maps cl_λ, $\lambda \in I$, yield a section

$$\mathrm{cl} = \mathrm{cl}_{M,U} : \Gamma(U, \Delta_\bullet) \to \Gamma(M, \Delta_\bullet).$$

If $c \in \Gamma(U, \Delta_\bullet)$, then we call $\mathrm{cl}(c)$ the *closure* of c in M. Obviously we have $\mathrm{cl}(c)|U = c$.
But note that, in general, cl is not a chain map.
More generally we may define the closure map $\mathrm{cl}_{M,A} : \Gamma(A, \Delta_\bullet(A)) \to \Gamma(M, \Delta_\bullet(M))$ for every locally closed set $A \in \mathcal{T}(M)$. We set $\mathrm{cl}_{M,A} = i_{M,\bar{A}} \circ \mathrm{cl}_{\bar{A},A}$ where $i_{M,\bar{A}} : \Gamma(\bar{A}, \Delta_\bullet(\bar{A})) = \Gamma_{\bar{A}}(M, \Delta_\bullet(M)) \to \Gamma(M, \Delta_\bullet(M))$ is the inclusion map and $\mathrm{cl}_{\bar{A},A}$ is defined as before (N.B. A is open in $\bar{A}$).

From Remark 4.13 we immediately obtain

Proposition 4.14. $\Delta_k(M,\Lambda)$ is a sa-flabby sheaf.

Since sd is a chain map (Lemma 4.1) the boundary maps $C_k(U,\varphi) \xrightarrow{\partial} C_{k-1}(U,\varphi)$ induce homomorphisms of sheaves

$$\partial : \Delta_k \to \Delta_{k-1}$$

with $\partial \circ \partial = 0$. Let $\Delta_k = 0$ for $k < 0$. Then $\Delta_\bullet = \{(\Delta_k | k \in \mathbf{Z}), \partial\}$ is a chain complex of sa-flabby sheaves.

Remark 4.15. Suppose that M is paracompact. Let $U \in \mathring{\mathcal{T}}(M)$. Since the homology functor commutes with direct limits and sd is a chain map we see that the homomorphism

$$H_*(U,\Lambda) \xrightarrow{\sim} H_*(C_\bullet(U,\varphi;\Lambda)) \to H_*(\Gamma(U,\Delta_\bullet))$$

is an isomorphism. Here we have chosen a triangulation φ of U. But the isomorphism $H_*(U,\Lambda) \xrightarrow{\sim} H_*(\Gamma(U,\Delta_\bullet))$ does not depend on this choice. Hence it is canonical.

Remark 4.16. Let $[c] \in \Gamma(M,\Delta_K)$ be a section represented by some $c \in C_k(M,\varphi)$. Then the support of the section $[c]$ of Δ_k (cf. II, §1) is a locally semialgebraic subset of M and coincides with the support of c defined above (Def. 1). This follows from Lemma 4.7.

§5 - Semialgebraic Borel-Moore homology with arbitrary coefficients and supports

In this section we extend the definition of semialgebraic Borel-Moore-homology to arbitrary locally complete spaces over R, arbitrary coefficients sheaves and arbitrary supports. Let M be a locally complete space over R and $\Delta_\bullet = \Delta_\bullet(M,\Lambda)$ be the chain complex of sheaves of simplicial chains on M defined in §4.
Let $\mathcal{F}$ be a sheaf of Λ-modules on M and Φ be a family of supports on M.

Definition 1. The group

$$H_k^\Phi(M,\mathcal{F}) := H_k(\Gamma_\Phi(M,\Delta_\bullet \otimes \mathcal{F}))$$

is called the k-th *(semialgebraic) Borel-Moore-homology group of M with coefficients in $\mathcal{F}$ and supports in Φ*. If $\Phi = cld(M)$, then we omit Φ in our notation.

Remark 5.1. If M is semialgebraic and $\varphi: X \xrightarrow{\sim} M$ is a triangulation, then $C_\bullet(M,\varphi;\Lambda) = C_\bullet(M,\varphi;\mathbf{Z}) \otimes_{\mathbf{Z}} \Lambda$. Hence $\Delta_\bullet(M,\Lambda) = \Delta_\bullet(M,\mathbf{Z}) \otimes_{\mathbf{Z}} \Lambda_M$ and $\Delta_\bullet(M,\Lambda) \otimes_\Lambda \mathcal{F} = \Delta_\bullet(M,\mathbf{Z}) \otimes_{\mathbf{Z}} \Lambda_M \otimes_\Lambda \mathcal{F} = \Delta_\bullet(M,\mathbf{Z}) \otimes_{\mathbf{Z}} \mathcal{F}$. We see that the groups $H_k^\Phi(M,\mathcal{F})$ defined above do not depend on the ground ring Λ. So we could choose $\mathbf{Z}$ as ground ring from the beginning. But it turns out later that it is quite useful to take an arbitrary principal ideal domain as ground ring (cf. e.g. Thm. 11.10).

Remark 5.2. If M is paracompact, then the groups $H_k(M,\Lambda_M)$ coincide with the groups $H_k(M,\Lambda)$ defined in §2 (Remark 4.15).

Remark 5.3. Let $\mathcal{G}$ be a sheaf on M. Then $\mathcal{G}$ is flat (i.e. every stalk $\mathcal{G}_x, x \in \tilde{M}$, is a flat Λ-module) if and only if $\mathcal{G}$ is torsion free (i.e. every stalk $\mathcal{G}_x, x \in \tilde{M}$, is a torsion free Λ-module) since Λ is a principal ideal domain. Moreover, $\mathcal{G}$ is torsion free if and only if $\Gamma(U,\mathcal{G})$ is torsion free for every $U \in \mathring{T}(M)$. For example, the sheaves $\Delta_k = \Delta_k(M,\Lambda)$ are torsion free.

Remark 5.4. The sheaves Δ_k are sa-flabby (Proposition 4.14). If Φ is paracompactifying, then $\Delta_k \otimes \mathcal{F}$ is Φ-soft by (II.4.8) and (II.4.22) and hence Φ-acyclic by (II.4.9) for every sheaf $\mathcal{F}$ on M. If M is of type (L) (i.e. every connected component of M is Lindelöf) and $\mathcal{F}$ is locally constant, then $\Delta_k \otimes \mathcal{F}$ is sa-flabby by (II.4.23) and hence Φ-acyclic for every family of supports on M by (II.4.10).

Lemma 5.5. Suppose M is semialgebraic and $\mathcal{G}$ is a torsion free sheaf on M. Let G be a Λ-module. Assume either that $\mathcal{G}$ is sa-flabby or that $\mathcal{G}$ is Φ-soft and Φ is paracompactifying. Then

$$\Gamma_\Phi(M,\mathcal{G} \otimes G_M) = \Gamma_\Phi(M,\mathcal{G}) \otimes G.$$

Proof. Since M is semialgebraic, we have $\Gamma_\Phi(M,\mathcal{G} \otimes G_M) = \varinjlim \Gamma_\Phi(M,\mathcal{G} \otimes G'_M)$ where G' runs through the finitely generated submodules of G (II.4.21). Hence we may assume that G is finitely generated. If G is free, our claim is evident. In general we choose a free resolution $0 \to G_1 \to G_0 \to G \to 0$ of G. Since $\mathcal{G}$ is torsion free, the sequence $0 \to \mathcal{G} \otimes G_1 \to \mathcal{G} \otimes G_0 \to \mathcal{G} \otimes G \to 0$ is also exact. Our hypotheses guarantee that $\mathcal{G} \otimes G_1$ is Φ-acyclic (Remark 5.4). Hence the sequence

(1) $\quad 0 \to \Gamma_\Phi(M, \mathcal{G} \otimes G_1) \to \Gamma_\Phi(M, \mathcal{G} \otimes G_0) \to \Gamma_\Phi(M, \mathcal{G} \otimes G) \to 0$

is exact. Since $\Gamma_\Phi(M, \mathcal{G})$ is torsion free, the sequence

(2) $\quad 0 \to \Gamma_\Phi(M, \mathcal{G}) \otimes G_1 \to \Gamma_\Phi(M, \mathcal{G}) \otimes G_0 \to \Gamma_\Phi(M, \mathcal{G}) \otimes G \to 0$

is exact. We already know that $\Gamma_\Phi(M, \mathcal{G} \otimes G_i) = \Gamma_\Phi(M, \mathcal{G}) \otimes G_i \quad (i = 0, 1)$. Comparing (1) and (2) we obtain the desired result.

For later use we mention the following

Corollary 5.6. Let $\mathcal{G}$ be a c-soft torsion free sheaf on M and G be a Λ-module. Then

$$\Gamma_c(M, \mathcal{G} \otimes G_M) = \Gamma_c(M, \mathcal{G}) \otimes G.$$

Proof. The family c of complete semialgebraic subsets of M is paracompactifying (Example II.1.7.iv)). Since each member K of c is semialgebraic, we have $\Gamma_c(M, \mathcal{G}') = \varinjlim \Gamma_c(M', \mathcal{G}')$ where $\mathcal{G}'$ is any sheaf on M and M' is running through the open semialgebraic subsets of M. Thus we may assume that M is semialgebraic. Now Lemma 5.5 applies.

Proposition 5.7. Let $\mathcal{F}$ be a locally constant sheaf on M. Then there are canonical isomorphisms

$$\Gamma_\Phi(M, \Delta_k(M, \Lambda) \otimes \mathcal{F}) \cong \varinjlim_{A \in \Phi} \Gamma(A, \Delta_k(A, \Lambda) \otimes \mathcal{F} \mid A).$$

In particular we have a canonical isomorphism

$$H_*^\Phi(M, \mathcal{F}) \cong \varinjlim_{A \in \Phi} H_*(A, \mathcal{F} \mid A).$$

Proof. We may assume that $\Phi = cld(A), A \in \bar{\mathcal{T}}(M)$. Let $(U_i \mid i \in I) \in \mathrm{Cov}(M)$ be a covering of M by open semialgebraic subsets such that $\mathcal{F} \mid U_i$ is constant for every $i \in I$. Since the sequences

$$0 \to \Gamma_A(M, \mathcal{G}) \to \prod_i \Gamma_{A \cap U_i}(U_i, \mathcal{G}) \rightrightarrows \prod_{i,j} \Gamma_{A \cap U_i \cap U_j}(U_i \cap U_j, \mathcal{G})$$

and

$$0 \to \Gamma(A, \mathcal{G}') \to \prod_i \Gamma(U_i \cap A, \mathcal{G}') \rightrightarrows \prod_{i,j} \Gamma(U_i \cap U_j \cap A, \mathcal{G}')$$

are exact for sheaves $\mathcal{G}$ and $\mathcal{G}'$ on M and A, we immediately reduce to the case where M is semialgebraic and $\mathcal{F} = G_M$, G a Λ-module.

By Prop. 4.12, Prop. 4.14, Remark 5.4 and Lemma 5.5 we conclude that

$\Gamma_A(M, \Delta_k(M, \Lambda) \otimes \mathcal{F}) = \Gamma_A(M, \Delta_k(M, \Lambda)) \otimes G = \Gamma(A, \Delta_k(A, \Lambda)) \otimes G =$
$\Gamma(A, \Delta_k(A, \Lambda) \otimes \mathcal{F} \mid A)$.

Our claim is proved.

In the most important cases we can interpret the Borel-Moore-homology as hypercohomology (cf. II, §10). We set

$$\Delta^k := \Delta^k(M,\Lambda) := \Delta_{-k}(M,\Lambda) \quad (k \in \mathbf{Z}).$$

Then $\Delta^\bullet$ is a cochain complex of sheaves of Λ-modules.

Proposition 5.8. Assume that at least one of the following conditions is satisfied:

1) M is of type (L), $\dim M < \infty$ and $\mathcal{F}$ is locally constant.
2) Φ is paracompactifying and $\dim \Phi < \infty$.

Then

$$H_k^\Phi(M,\mathcal{F}) = \mathsf{H}_\Phi^{-k}(M, \Delta^\bullet \otimes \mathcal{F})$$

for every $k \geq 0$.

Proof. The hypercohomology is defined in both cases (cf. II, §10). By Remark 5.4 the sheaves $\Delta^k \otimes \mathcal{F}$ are Φ-acyclic. Since the hypercohomology may be computed by means of arbitrary Φ-acyclic resolutions (cf. II, §10) we see that

$$\mathsf{H}_\Phi^{-k}(M,\Delta^\bullet \otimes \mathcal{F}) = H^{-k}(\Gamma_\Phi(M,\Delta^\bullet \otimes \mathcal{F})) =$$
$$H_k(\Gamma_\Phi(M,\Delta_\bullet \otimes \mathcal{F})) = H_k^\Phi(M,\mathcal{F}).$$

Obviously the homology groups $H_*^\Phi(M,\mathcal{F})$ are functorial in $\mathcal{F}$. Short exact sequences of coefficient sheaves yield long exact homology sequences:

Proposition 5.9. Let $0 \to \mathcal{F}' \to \mathcal{F} \to \mathcal{F}'' \to 0$ be an exact sequence of sheaves on M. Suppose either that Φ is paracompactifying or that M is of type (L) and $\mathcal{F}'$ is locally constant. Then there is a natural long exact sequence

$$\ldots \to H_k^\Phi(M,\mathcal{F}') \to H_k^\Phi(M,\mathcal{F}) \to H_k^\Phi(M,\mathcal{F}'') \xrightarrow{\partial} H_{k-1}^\Phi(M,\mathcal{F}') \to \ldots.$$

Proof. By hypothesis $\Delta_k \otimes \mathcal{F}'$ is Φ-acyclic for every k (cf. Remark 5.4). Hence the sequence of chain complexes $0 \to \Gamma_\Phi(\Delta_\bullet \otimes \mathcal{F}') \to \Gamma_\Phi(\Delta_\bullet \otimes \mathcal{F}) \to \Gamma_\Phi(\Delta_\bullet \otimes \mathcal{F}'') \to 0$ is exact and yields the long exact sequence.

Remark 5.10. Suppose Φ is paracompactifying. Let $n \in \mathbf{N}$. Proposition 5.9 says that the system of functors

$$\mathcal{F} \mapsto H_{n-i}^\Phi(M,\mathcal{F}), \quad i \in \mathbf{Z},$$

together with the connecting homomorphisms $H_{n-i}^\Phi(M,\mathcal{F}'') \to H_{n-i-1}^\Phi(M,\mathcal{F}')$ (for short exact sequences $0 \to \mathcal{F}' \to \mathcal{F} \to \mathcal{F}'' \to 0$) is an exact connected sequence of functors in the sense of [B, II.6] (= an exact ∂-functor in the sense of [Gr]).

If $U \in \mathring{\mathcal{T}}(M)$, then the restriction map $\Gamma_\Phi(M,\Delta_\bullet \otimes \mathcal{F}) \to \Gamma_{\Phi\cap U}(U,\Delta_\bullet \otimes \mathcal{F})$ yields a restriction map

$$H_*^\Phi(M,\mathcal{F}) \xrightarrow{\text{res}} H_*^{\Phi\cap U}(U,\mathcal{F}).$$

In particular we see that

$$U \mapsto H_k(U, \mathcal{F}) \quad (U \in \dot{\mathcal{T}}(M))$$

is a presheaf on M. Its associated sheaf is denoted by $\mathcal{H}_k(M, \mathcal{F})$. If $\mathcal{F} = \Lambda_M$ we often simply write $\mathcal{H}_k$ or $\mathcal{H}_k(M)$ instead of $\mathcal{H}_k(M, \Lambda) := \mathcal{H}_k(M, \Lambda_M)$. In the special case where M is paracompact we already considered these restriction maps and homology sheaves in §2. Obviously the sheaves $\mathcal{H}_k(M, \mathcal{F})$ are just the homology sheaves of the chain complex of sheaves $\Delta_\bullet \otimes \mathcal{F}$.

Proposition 5.11. Let $U \in \dot{\mathcal{T}}(M)$ and $A := M \setminus U$. Suppose that M is of type (L) and $\mathcal{F}$ is locally constant. Then there is an exact sequence
$\ldots \to H_k^{\Phi|A}(A, \mathcal{F} \mid A) \to H_k^{\Phi}(M, \mathcal{F}) \xrightarrow{\text{res}} H_k^{\Phi \cap U}(U, \mathcal{F}) \to H_{k-1}^{\Phi|A}(A, \mathcal{F}) \to \ldots$
which is functorial in $\mathcal{F}$.

Proof. Since $\Delta_k(M, \Lambda) \otimes \mathcal{F}$ is sa-flabby (Remark 5.4), the sequence of chain complexes
$0 \to \Gamma_{\Phi|A}(M, \Delta_\bullet \otimes \mathcal{F}) \to \Gamma_\Phi(M, \Delta_\bullet \otimes \mathcal{F}) \to \Gamma_{\Phi \cap U}(U, \Delta_\bullet \otimes \mathcal{F}) \to 0$
is exact and hence yields a long exact sequence
$\ldots \to H_k^{\Phi|A}(M, \mathcal{F}) \to H_k^{\Phi}(M, \mathcal{F}) \to H_k^{\Phi \cap U}(U, \mathcal{F}) \to \ldots.$
From Prop. 5.7 we know that $H_*^{\Phi|A}(M, \mathcal{F}) = H_*^{\Phi|A}(A, \mathcal{F} \mid A)$.

The long exact sequence in 5.11 is a more general version of the long exact sequence (2.8).

Finally we generalize Prop. 2.14.

Proposition 5.12. Suppose that dim $M \leq n$ and $\mathcal{F}$ is an arbitrary sheaf on M. Then

$$\Gamma(U, \mathcal{H}_n(M, \mathcal{F})) = H_n(U, \mathcal{F})$$

for every $U \in \dot{\mathcal{T}}(M)$.

Proof. The chain complex $\Delta_\bullet \otimes \mathcal{F}$ has the form

$$0 \to \Delta_n \otimes \mathcal{F} \to \Delta_{n-1} \otimes \mathcal{F} \to \ldots.$$

Hence we have
$\Gamma(U, \mathcal{H}_n(M, \mathcal{F})) = \Gamma(U, \mathrm{Ker}\,(\Delta_n \otimes \mathcal{F} \to \Delta_{n-1} \otimes \mathcal{F})) =$
$\mathrm{Ker}\,(\Gamma(U, \Delta_n \otimes \mathcal{F}) \to \Gamma(U, \Delta_{n-1} \otimes \mathcal{F})) = H_n(U, \mathcal{F})$.

Remark 5.13. The results on homology-manifolds and fundamental classes we obtained in §3 under the hypothesis that the spaces are paracompact remain true in general.

§6 - Induced maps

We consider a locally semialgebraic map $f: M \to N$ between locally complete spaces M and N over R and study the following question: When does f induce a homomorphism in Borel-Moore-homology?

We work with the functor $f_!$ („direct image with proper support") defined in II, §8.

Lemma 6.1. Let $\mathcal{F}$ be a *sa*-flabby torsion free sheaf on M and $\mathcal{G}$ be a sheaf on N. Then the canonical homomorphism

$$f_!\mathcal{F} \otimes \mathcal{G} \to f_!(\mathcal{F} \otimes f^*\mathcal{G})$$

is an isomorphism.

Proof. We verify the claim in the stalks on $\tilde{N}$. So let $y \in \tilde{N}$. Then $\tilde{f}^{-1}(y)$ is a locally complete space over $k(y)$ (Cor. II.8.3). From Cor. II.8.8 we know that
$f_!(\mathcal{F} \otimes f^*\mathcal{G})_y = \Gamma_c(\tilde{f}^{-1}(y), (\mathcal{F} \mid \tilde{f}^{-1}(y)) \otimes \mathcal{G}_y)$ and
$(f_!\mathcal{F} \otimes \mathcal{G})_y = \Gamma_c(\tilde{f}^{-1}(y), \mathcal{F}) \otimes \mathcal{G}_y$. Since $\mathcal{F} \mid \tilde{f}^{-1}(y)$ is torsion free and c-soft by Prop. II.4.8, Cor. 5.6 gives the result.

Now we define a canonical homomorphism of complexes of sheaves

$$\alpha : f_!\Delta_\bullet(M, \Lambda) \to \Delta_\bullet(N, \Lambda).$$

(The coefficients Λ will be omitted in the following). For any section $c \in \Delta_k$ we denote the support of c by $\mid c \mid$. We assume for a while that N is semialgebraic. Let $c \in \Gamma(N, f_!\Delta_n(M))$. Then $c \in \Gamma(M, \Delta_n(M))$ and the restriction $f \mid \mid c \mid : \mid c \mid \to N$ of f to $\mid c \mid$ is proper and semialgebraic. In particular $\mid c \mid$ is a semialgebraic set. We have $\dim f(\mid c \mid) \leq n$ and $\dim f(\mid \partial c \mid) \leq n - 1$. Hence the closure map (cf. Remark 4.13) yields a canonical isomorphism

$$\Gamma(f(\mid c \mid), \Delta_n(f(\mid c \mid))) = \Gamma(f(\mid c \mid) \setminus f(\mid \partial c \mid), \Delta_n(f(\mid c \mid))).$$

Notation. For any locally complete space P we denote the cycles (resp. boundaries) of $\Gamma(P, \Delta_\bullet(P))$ by $Z_\bullet(P)$ (resp. $B_\bullet(P)$).

In §2 we defined an induced homomorphism

$$\begin{array}{ccc} f_* : H_*(\mid c \mid, \mid \partial c \mid) & \longrightarrow & H_*(f(\mid c \mid), f(\mid \partial c \mid)) \\ \| & & \| \\ H_*(\mid c \mid \setminus \mid \partial c \mid) & & H_*(f(\mid c \mid) \setminus f(\mid \partial c \mid)). \end{array}$$

Since $\dim \mid c \mid \leq n$ and $\dim f(\mid c \mid) \leq n$, we have $H_n(\mid c \mid \setminus \mid \partial c \mid) = Z_n(\mid c \mid \setminus \mid \partial c \mid)$ and $H_n(f(\mid c \mid) \setminus f(\mid \partial c \mid)) = Z_n(f(\mid c \mid) \setminus f(\mid \partial c \mid))$. Therefore we have a map $f_* : Z_n(\mid c \mid \setminus \mid \partial c \mid) \to Z_n(f(\mid c \mid) \setminus f(\mid \partial c \mid))$. Let $cl : Z_n(f(\mid c \mid) \setminus f(\mid \partial c \mid)) \hookrightarrow \Gamma(f(\mid c \mid), \Delta_n(f(\mid c \mid))) \hookrightarrow \Gamma(N, \Delta_n(N))$ be the closure map (Remark 4.13).
Obviously c yields a cycle in $Z_n(\mid c \mid) \setminus \mid \partial c \mid)$. We denote this cycle by $\mathring{c}$. Now we define

$$\alpha_N(c) := cl \circ f_*(\mathring{c}).$$

In this way we obtain a homomorphism (of Λ-modules) $\alpha_N : \Gamma(N, f_! \Delta_\bullet(M, \Lambda)) \to \Gamma(N, \Delta_\bullet(N, \Lambda))$.

Lemma 6.2. α_N is a chain map.

Proof. Consider the diagram

$$\begin{array}{ccc}
\Gamma(M, \Delta_n(M)) & \xrightarrow{\partial} & \Gamma(M, \Delta_{n-1}(M)) \\
\uparrow cl & & \uparrow cl \\
H_n(|c| \setminus |\partial c|) = Z_n(|c| \setminus |\partial c|) & \xrightarrow{\partial} & Z_{n-1}(|\partial c|) = H_{n-1}(|\partial c|) \\
\downarrow f_* & & \downarrow f_* \\
H_n(f(|c|) \setminus f(|\partial c|)) = Z_n(f(|c|) \setminus f(|\partial c|)) & \xrightarrow{\partial} & Z_{n-1}(f(|\partial c|)) = H_{n-1}(f(|\partial c|)) \\
\downarrow cl & & \downarrow cl \\
\Gamma(N, \Delta_n(N)) & \xrightarrow{\partial} & \Gamma(N, \Delta_{n-1}(N))
\end{array}$$

The second { resp. third } horizontal arrow is the boundary map in the long exact sequence (2.8) of the closed pair $(|c|, |\partial c|)$ { resp. $(f(|c|), f(|\partial c|))$}. The diagram commutes (cf. §2). But this just means that α_N is a chain map.

Now we consider the family of chain maps

$$\begin{array}{ccc}
\alpha_U : \Gamma(U, f_! \Delta_\bullet(f^{-1}(U))) & \longrightarrow & \Gamma(U, \Delta_\bullet(U)) \\
\| & & \| \\
\Gamma(U, f_! \Delta_\bullet(M)) & & \Gamma(U, \Delta_\bullet(N))
\end{array}$$

where U runs through $\mathring{\gamma}(N)$. The diagram

$$\begin{array}{ccc}
\Gamma(U, f_! \Delta_\bullet(M)) & \xrightarrow{\alpha_U} & \Gamma(U, \Delta_\bullet(N)) \\
\downarrow \text{res} & & \downarrow \text{res} \\
\Gamma(V, f_! \Lambda_\bullet(M)) & \xrightarrow{\alpha_V} & \Gamma(V, \Delta_\bullet(N))
\end{array}$$

commutes for $V \subset U$. This implies:

1) $\alpha = (\alpha_U)_{U \in \mathring{\gamma}(N)}$ is a homomorphism of complexes of sheaves $f_! \Delta_\bullet(M) \to \Delta_\bullet(N)$.

2) The definition extends to the case where N is an arbitrary (not necessarily semialgebraic) space. (Choose some admissible covering of N by open semialgebraic subsets!)

Now let $\mathcal{F}$ be a sheaf on M, $\mathcal{G}$ be a sheaf on N and $\delta : \mathcal{F} \to f^* \mathcal{G}$ be a sheaf homomorphism. Such a pair (f, δ) is called a *covariant map between the pairs* $(M, \mathcal{F})$ and $(N, \mathcal{G})$ (cf. II, §2). From Lemma 6.1 we know that

$$f_!(\Delta_\bullet(M) \otimes f^* \mathcal{G}) = f_! \Delta_\bullet(M) \otimes \mathcal{G}$$

and we obtain a homomorphism

$$f_!(\Delta_\bullet(M) \otimes \mathcal{F}) \xrightarrow{\delta_\bullet} f_!(\Delta_\bullet(M) \otimes f^* \mathcal{G}) \xrightarrow{\alpha \otimes \text{id}} \Delta_\bullet(N) \otimes \mathcal{G}$$

of complexes of sheaves. If Φ and Ψ are families of supports on M and N such that $f \mid A : A \to N$ is proper and semialgebraic for every $A \in \Phi$ (we say that „*f is proper and semialgebraic on* Φ") and $f(\Phi) \subset \Psi$ (i.e. $f(A) \in \Psi$ for every $A \in \Phi$), then this homomorphism of complexes of sheaves induces a chain map

$$(f,\delta)_\bullet : \Gamma_\Phi(M, \Delta_\bullet(M) \otimes \mathcal{F}) \to \Gamma_\Psi(N, \Delta_\bullet(N) \otimes \mathcal{G})$$

and hence a homomorphism

$$(f,\delta)_* : H_*^\Phi(M,\mathcal{F}) \to H_*^\Psi(N,\mathcal{G})$$

in the Borel-Moore-homology.
If $\mathcal{F} = f^*\mathcal{G}$ and $\delta = \mathrm{id}$, we simply write f_* instead of $(f,\delta)_*$. Let $(g,\varepsilon) : (N,\mathcal{G}) \to (L,\mathcal{H})$ be another covariant map. Then we may compose (f,δ) and $(g,\varepsilon) : (g,\varepsilon) \circ (f,\delta) := (g \circ f, f^*(\varepsilon) \circ \delta)$. (N.B. $\mathcal{F} \xrightarrow{\delta} f^*\mathcal{G} \xrightarrow{f^*(\varepsilon)} f^*g^*\mathcal{H} = (g \circ f)^*\mathcal{H}$).
Let Θ be a family of supports on L and suppose that g is proper and semialgebraic on Ψ and $g(\Psi) \subset \Theta$. Then we have

$$((g,\varepsilon) \circ (f,\delta))_\bullet = (g,\varepsilon)_\bullet \circ (f,\delta)_\bullet$$

and

$$((g,\varepsilon) \circ (f,\delta))_* = (g,\varepsilon)_* \circ (f,\delta)_*.$$

This is an easy consequence of the definition.

Examples 6.3. Let $f : M \to N$ be a locally semialgebraic map.
a) f induces a homomorphism $f_* : H_*^c(M, f^*\mathcal{G}) \to H_*^c(N,\mathcal{G})$ in the homology with complete supports.
b) If f is proper and semialgebraic, then f induces a homomorphism $f_* : H_*(M, f^*\mathcal{G}) \to H_*(N,\mathcal{G})$ in the homology with closed supports.
c) If f is partially proper, then f induces a homomorphism $f_* : H_*^{sa}(M, f^*\mathcal{G}) \to H_*^{sa}(N,\mathcal{G})$ in the homology with semialgebraic supports.
d) If f is semialgebraic, then f induces a homomorphism $f_* : H_*^{pc}(M, f^*\mathcal{G}) \to H_*^{pc}(N,\mathcal{G})$ in the homology with partially complete supports.

The statements a), b), c) are clear. Notice in b) that every proper map $f : M \to N$ is necessarily semialgebraic if M is of type (L) ([DK$_3$, I.5.9]). Statement d) follows from [DK$_3$, I.6.2.iii), I.6.12]).

§7 - Homotopy

The unit interval [0,1] over R is denoted by I. We consider a locally complete space M and the closed embeddings

$$i_t : M \to M \times I, x \mapsto (x,t) \quad (t \in I).$$

Let $\pi : M \times I \to M$ be the projection. The chain maps

$$(i_t)_\bullet : \Gamma(U, \Delta_\bullet(M)) \to \Gamma(\pi^{-1}(U), \Delta_\bullet(M \times I))$$

defined in §6 yields homomorphisms of sheaf complexes $(i_t)_\bullet : \Delta_\bullet(M) \to \pi_*\Delta_\bullet(M \times I)$. (Again we omit the ground ring Λ in our notation). We will prove

Theorem 7.1. The homomorphisms $(i_0)_\bullet$ and $(i_1)_\bullet$ are (chain) homotopic.

We construct a chain homotopy D from $(i_0)_\bullet$ to $(i_1)_\bullet$. The construction shows that it suffices to do this when M is semialgebraic. So we now assume that M is semialgebraic.
Let $c \in \Gamma(M, \Delta_n(M))$. Again we denote the support of c by $|\, c \,|$. Let $|\, \mathring{c} \,| := |\, c \,| \setminus |\, \partial c \,|$ and $\mathring{c} \in \Gamma(|\, \mathring{c} \,|, \Delta_n(|\, c \,|))$ be the restriction of c to $|\, \mathring{c} \,|$. Since $\dim \,|\, c \,| = n$, we have $H_n(|\, \mathring{c} \,|) = Z_n(|\, \mathring{c} \,|)$. Note that $\mathring{c} \in Z_n(|\, \mathring{c} \,|)$. Let $\partial I = \{0,1\}$ be the boundary of I and $\mathring{I} = I \setminus \partial I$. We consider the long exact sequence (2.8) of the closed pair $(|\, \mathring{c} \,| \times I, |\, \mathring{c} \,| \times \partial I)$
$\ldots \to H_{n+1}(|\, \mathring{c} \,| \times I) \xrightarrow{\text{res}} H_{n+1}(|\, \mathring{c} \,| \times \mathring{I}) \xrightarrow{\partial} H_n(|\, \mathring{c} \,| \times \partial I) \xrightarrow{i_*} H_n(|\, \mathring{c} \,| \times I) \to \ldots .$
Here i denotes the inclusion $|\, \mathring{c} \,| \times \partial I \hookrightarrow |\, \mathring{c} \,| \times I$ and i_* the induced map in homology. Since $|\, \mathring{c} \,|$ is a proper deformation retract of $|\, \mathring{c} \,| \times I$, we conclude from Theorem 2.10 that $H_{n+1}(|\, \mathring{c} \,| \times I) = H_{n+1}(|\, \mathring{c} \,|) = 0$. Furthermore we have $H_{n+1}(|\, \mathring{c} \,| \times \mathring{I}) = Z_n(|\, \mathring{c} \,| \times \mathring{I})$ and $H_n(|\, \mathring{c} \,| \times \partial I) = Z_n(|\, \mathring{c} \,| \times \partial I)$. Thus our sequence has the form
$0 \to Z_{n+1}(|\, \mathring{c} \,| \times \mathring{I}) \xrightarrow{\partial} Z_n(|\, \mathring{c} \,| \times \partial I) \xrightarrow{I_*} H_n(|\, \mathring{c} \,| \times I) \to \ldots .$
Let $\mathring{c} \times 0$ (resp. $\mathring{c} \times 1$) be the cycle in $Z_n(|\, \mathring{c} \,| \times \{0\}) \subset Z_n(|\, \mathring{c} \,| \times \partial I)$ (resp. $Z_n(|\, \mathring{c} \,| \times \{1\}) \subset Z_n(|\, \mathring{c} \,| \times \partial I)$) given by $\mathring{c}$. We observe that $i_*(\mathring{c} \times 0) = (i_0)_*(\mathring{c})$ and $i_*(\mathring{c} \times 1) = (i_1)_*(\mathring{c})$. Here $(i_t)_*$ denotes the homomorphism $H_*(|\, \mathring{c} \,|) \to H_*(|\, \mathring{c} \,| \times I)$ induced by i_t. From Theorem 2.10 we conclude that $i_*(\mathring{c} \times 0) = i_*(\mathring{c} \times 1)$. Hence there is a uniquely determined cycle $D'_M(c) \in Z_{n+1}(|\, \mathring{c} \,| \times \mathring{I})$ with $\partial D'_M(c) = \mathring{c} \times 1 - \mathring{c} \times 0$.
Let $\mathrm{cl} = \mathrm{cl}_{M \times I, |\mathring{c}| \times \mathring{I}}$ be the closure map (Remark 4.13) and define

$$D_M(c) := \mathrm{cl}\,(D'_M(c)) \in \Gamma(M \times I, \Delta_{n+1}(M \times I)).$$

Obviously we obtain a linear map

$$D_M : \Gamma(M, \Delta_n(M)) \to \Gamma(M \times I, \Delta_{n+1}(M \times I))$$

in this way (for every $n \geq 0$).
We claim that D_M is a chain homotopy from $(i_0)_\bullet$ to $(i_1)_\bullet$.

Lemma 7.2. We have

$$\partial \circ D_M(c) + D_M(\partial c) = (i_1)_\bullet(c) - (i_0)_\bullet(c)$$

for every $c \in \Gamma(M, \Delta_n(M))$.

Proof. We consider the diagram

$$\begin{array}{ccc} Z_{n+1}(|\,\mathring{c}\,|\times \mathring{I}) & \xrightarrow{\partial_1} & Z_n(|\,\partial c\,|\times \mathring{I}) \\ \partial_4 \downarrow & & \downarrow \partial_2 \\ Z_n(|\,\mathring{c}\,|\times \partial I) & \xrightarrow[\partial_3]{} & Z_{n-1}(|\,\partial c\,|\times \partial I). \end{array}$$

Here ∂_1 {resp. $\partial_2, \partial_3, \partial_4$} is the boundary map in the long exact sequence (2.8) of the pair $(|\,c\,|\times \mathring{I}, |\,\partial c\,|\times \mathring{I})$ {resp. $(|\,\partial c\,|\times I, |\,\partial c\,|\times \partial I)$, $(|\,c\,|\times \partial I, |\,\partial c\,|\times \partial I)$, $(|\,\mathring{c}\,|\times I, |\,\mathring{c}\,|\times \partial I)$}. It is easily seen that the diagram commutes. Since $\partial_3(\mathring{c}\times 1 - \mathring{c}\times 0) = \partial c\times 1 - \partial c \times 0$, we see by definition of D'_M that

$$\partial_1 D'_M(c) = -D'_M(\partial c).$$

We denote the boundary in $\Gamma(|\,\mathring{c}\,|\times I, \Delta_\bullet(|\,\mathring{c}\,|\times I))$ { resp. $\Gamma(|\,c\,|\times \mathring{I}, \Delta_\bullet(|\,c\,|\times \mathring{I}))$} by $\partial_{|\mathring{c}|\times I}$ { resp. $\partial_{|c|\times \mathring{I}}$} and compute

$\partial D_M(c) = \partial \mathrm{cl}_{M\times I, |\mathring{c}|\times \mathring{I}}(D'_M(c))$
$= \mathrm{cl}_{M\times I, |\mathring{c}|\times I}(\partial_{|\mathring{c}|\times I}\mathrm{cl}_{|\mathring{c}|\times I, |\mathring{c}|\times \mathring{I}}(D'_M(c)))+$
$\mathrm{cl}_{M\times I, |c|\times \mathring{I}}(\partial_{|c|\times \mathring{I}}(D'_M(c)))$
$= \mathrm{cl}_{M\times I, |\mathring{c}|\times \partial I}(\partial_4 D'_M(c)) + \mathrm{cl}_{M\times I, |\partial c|\times \mathring{I}}(\partial_1 D'_M(c))$
$= \mathrm{cl}_{M\times I, |\mathring{c}|\times \partial I}(\mathring{c}\times 1 - \mathring{c}\times 0) - \mathrm{cl}_{M\times I, |\partial c|\times \mathring{I}}(D'_M(\partial c))$
$= (i_1)_\bullet(c) - (i_0)_\bullet(c) - D_M(\partial c).$

Lemma 7.2 is proven.

We consider the family of homotopies

$$\begin{array}{ccc} D_U : \Gamma(U, \Delta_\bullet(U)) & \longrightarrow & \Gamma(U\times I, \Delta_{\bullet+1}(U\times I)) \\ \| & & \| \\ \Gamma(U, \Delta_\bullet(M)) & & \Gamma(U\times I, \Delta_{\bullet+1}(M\times I)) \end{array}$$

where U runs through $\mathring{\gamma}(M)$. It is obvious that the diagram

$$\begin{array}{ccc} \Gamma(U, \Delta_\bullet(M)) & \xrightarrow{D_U} & \Gamma(U\times I, \Delta_{\bullet+1}(M\times I)) \\ \downarrow \mathrm{res} & & \downarrow \mathrm{res} \\ \Gamma(V, \Delta_\bullet(M)) & \xrightarrow{D_V} & \Gamma(V\times I, \Delta_{\bullet+1}(M\times I)) \end{array}$$

commutes for $V \subseteq U$. This implies:

1) $D = (D_U)_{U\in\mathring{\gamma}(M)}$ is a chain homotopy between the maps $(i_0)_\bullet$ and $(i_1)_\bullet$ of complexes of sheaves.

2) If M is arbitrary and $(M_i \mid i\in I)$ is an admissible covering of M by open semialgebraic subsets, then the chain homotopies D_i defined on each M_i by the preceding construction glue together to give a chain homotopy D between the maps

$$(i_0)_\bullet, (i_1)_\bullet : \Delta_\bullet(M) \rightrightarrows \pi_*\Delta_\bullet(M\times I).$$

The proof of Theorem 7.1 is finished.

Now let $\mathcal{F}$ be a sheaf and Φ be a family of supports on M. The sheaf $\pi^*\mathcal{F}$ is denoted by $\mathcal{F}\times I$, the family $\pi^{-1}\Phi = \{A\in \bar{T}(M\times I) \mid \pi(A)\in\Phi\}$ by $\Phi\times I$. Notice that $i_t^*(\mathcal{F}\times I) = \mathcal{F}$ for every $t\in[0,1]$ and that $\pi_*(\Delta_\bullet(M\times I)\otimes(\mathcal{F}\times I)) = \pi_*(\Delta_\bullet(M\times I))\otimes\mathcal{F}$ by Lemma 6.1.

By Theorem 7.1 we derive

Corollary 7.3. The chain maps
$(i_0)_\bullet, (i_1)_\bullet : \Gamma_\Phi(M, \Delta_\bullet(M) \otimes \mathcal{F}) \to \Gamma_{\Phi\times I}(M \times I, \Delta_\bullet(M \times I) \otimes (\mathcal{F} \times I))$ are homotopic.

Since these maps induce the homomorphisms in homology, we obtain

Corollary 7.4. i_0 and i_1 induce the same homomorphism
$(i_0)_* = (i_1)_* : H_*^\Phi(M, \mathcal{F}) \to H_*^{\Phi\times I}(M \times I, \mathcal{F} \times I)$.

Now we consider (covariant) maps $(f, \delta) : (M, \mathcal{F}) \to (N, \mathcal{G})$ between pairs consisting of a (locally complete) space and a sheaf on this space (cf. §6). As before we omit δ if $\mathcal{F} = f^*\mathcal{G}$ and $\delta = \mathrm{id}$. Let Φ and Ψ be families of supports on M and N.

Definition 1. Two (covariant) maps $(f, \delta), (g, \varepsilon) : (M, \mathcal{F}) \rightrightarrows (N, \mathcal{G})$ are called *properly homotopic with respect to* Φ *and* Ψ (written $(f, \delta) \underset{\Phi,\Psi}{\simeq} (g, \varepsilon)$) if there exists a map $(F, \lambda) : (M \times I, \mathcal{F} \times I) \to (N, \mathcal{G})$ such that $(F, \lambda) \circ i_0 = (f, \delta), (F, \lambda) \circ i_1 = (g, \varepsilon)$, F is proper and semialgebraic on $\Phi \times I$ and $F(\Phi \times I) \subset \Psi$.
(NB: Then f and g are proper and semialgebraic on Φ and $f(\Phi) \subset \Psi$ and $g(\Phi) \subset \Psi$).

From Cor. 7.4 we immediately obtain the following „homotopy axiom" for semialgebraic Borel-Moore-homology.

Corollary 7.5. If $(f, \delta) \underset{\Phi,\Psi}{\simeq} (g, \varepsilon)$, then $(f, \delta)_* = (g, \varepsilon)_* : H_*^\Phi(M, \mathcal{F}) \to H_*^\Psi(N, \mathcal{G})$.

Examples 7.6. Let $f, g : M \rightrightarrows N$ be locally semialgebraic maps and $H : M \times I \to N$ be a homotopy from f to g, i.e. $H(-, 0) = f$ and $H(-, 1) = g$. Let G be a Λ-module.
a) We have

$$f_* = g_* : H_*^c(M, G) \to H_*^c(N, G).$$

b) If H is proper and semialgebraic (i.e. f and g are properly homotopic, cf. Def. 2 in §2) then

$$f_* = g_* : H_*(M, G) \to H_*(N, G).$$

c) If H is partially proper, then

$$f_* = g_* : H_*^{sa}(M, G) \to H_*^{sa}(N, G).$$

d) If H is semialgebraic, then

$$f_* = g_* : H_*^{pc}(M, G) \to H_*^{pc}(N, G).$$

Example 7.7. Let M be paracompact and $\varphi : X \xrightarrow{\sim} M$ be a good triangulation of M (c [DK$_1$, §2], [DK$_3$, III. §1]). Then the core M_0 of M with respect to φ is a strong deformatio retract of M and the deformation retraction F may be chosen as a semialgebraic map. Fror Example 7.6 d) we conclude that

$$H_*(M_0, G) = H_*^{pc}(M_0, G) \xrightarrow{\sim} H_*^{pc}(M, G).$$

ence $H_*^{pc}(M,G)$ is just the simplicial homology $H_*(C_\bullet(M_0,\varphi;G))$ of the closed abstract omplex $K(M_0,\varphi)$.
ow suppose that M is semialgebraic. Then $H_*^c(M,G) = H_*^{pc}(M,G) = I_*(C_\bullet(M_0,\varphi;G))$ and we see that the Borel-Moore-homology groups $H_k^c(M,G)$ with comlete supports are the homology groups we defined in [D], [DK$_1$, §3].

xample 7.8. We consider the situation of Example II.2.7 . So let M be a locally finite mplicial complex over R and A be a closed subcomplex of M. Let A_2 be the closure $\overline{t_{M''}(A)} \cap M$ of the open star neighbourhood $\mathrm{St}_{M''}(A)$ of A in the second barycentric ıbdivision M'' of M. There is a canonical strong deformation retraction

$$H : A_2 \times I \to A_2$$

rm A_2 to A explicitly described in [DK$_2$, §2]. We have $H(-,0) = \mathrm{id}_{A_2}$ and $H(-,1) = i \circ r$ ith $i : A \to A_2$ the inclusion map and r a retraction from A_2 to A. In fact H is simply the ap $(x,t) \mapsto (1-t)\cdot x + t \cdot r(x)$ (loc. cit.).
et $\mathcal{F}$ be a *locally constant* sheaf on M. H is the restriction of a deformation retraction

$$G : \mathrm{St}_{M'}(A) \times I \to \mathrm{St}_{M'}(A)$$

M' the first barycentric subdivision of M) which has the following property: If $\tau \subset A$ is open simplex of M', then $G(\mathrm{St}_{M'}(\tau) \times I) \subset \mathrm{St}_{M'}(\tau)$ (loc. cit.). This implies that $^*(\mathcal{F} \mid A_2) = \mathcal{F} \times I \mid A_2 \times I$. Moreover, the deformation retraction H is a proper mialgebraic map ([DK$_2$, 3.8]).
ow we conclude from Cor. 7.5 that $i_* \circ r_* : H_*(A_2, \mathcal{F} \mid A_2) \to H_*(A_2, \mathcal{F} \mid A_2)$ is the identity ap. Since $r \circ i = \mathrm{id}, r_* \circ i_* : H_*(A, \mathcal{F} \mid A) \to H_*(A, \mathcal{F} \mid A)$ is also the identity map. Hence

$$i_* : H_*(A, \mathcal{F} \mid A) \to H_*(A_2, \mathcal{F} \mid A_2)$$

an isomorphism.

analogy to Prop. II.2.9 we get

roposition 7.9. Let M be paracompact and $\mathfrak{A} \subset \bar{\mathcal{T}}(M)$ be a finite family which is osed under intersections. Then there are paracompactifying families $\Phi(A)$ of supports on $(A \in \mathfrak{A})$ such that $A \in \Phi(A), \Phi(A) \cap \Phi(B) = \Phi(A \cap B)$ for any $A, B \in \mathfrak{A}$ and the nonical maps

$$H_A^*(M, \mathcal{F}) \longrightarrow H_{\Phi(A)}^*(M, \mathcal{F})$$
$$H_*^A(M, \mathcal{F}) \longrightarrow H_*^{\Phi(A)}(M, \mathcal{F})$$

e isomorphisms for every locally constant sheaf $\mathcal{F}$ and every $A \in \mathfrak{A}$. If closed locally mialgebraic neighbourhoods $W(A)$ of A ($A \in \mathfrak{A}$) are given, then we may choose $\Phi(A)$ in ch a way that each member B of $\Phi(A)$ is contained in $W(A)$.

oof. We may assume that M is a locally finite simplicial complex and that the members $\mathfrak{A}$ and the neighbourhoods $W(A), A \in \mathfrak{A}$, are subcomplexes. Let $\Phi(A)$ be the family defined in the proof of Prop. II.2.9. Then $\Phi(A) \cap \Phi(B) = \Phi(A \cap B)$ and we have the claimed isomorphism in cohomology (cf. II.2.9). We also have the isomorphism in homology, since $H_*^A(M, \mathcal{F}) = H_*(A, \mathcal{F})$ and $H_*^{\Phi(A)}(M, \mathcal{F}) = \varinjlim_{B \in \Phi(A)} H_*(B, \mathcal{F})$ by Prop. 5.7 and $H_*(A_n, \mathcal{F}) \to H_*(A_{n+1}, \mathcal{F})$ (notation as in the proof of II.2.9) is an isomorphism by Example 7.8.

Remark 7.10. Consider the situation of Prop. 7.9. It follows from the statement that, given paracompactifying support families Ψ_A with $A \in \Psi_A$ $(A \in \mathfrak{A})$, we may choose $\Phi(A)$ as a subfamily of Ψ_A.

§8 - The cap product

We use the procedure described in [B, V.10] to define a cap product between cohomology and homology.

Let M be a locally complete space, $\mathcal{F}, \mathcal{G}$ be sheaves and Φ, Ψ be paracompactifying families of supports on M. By $\mathfrak{E}$ we denote the class of short exact sequences $0 \to \mathcal{H}' \to \mathcal{H} \to \mathcal{H}'' \to 0$ of sheaves on M such that $0 \to \mathcal{H}'_x \to \mathcal{H}_x \to \mathcal{H}''_x \to 0$ splits for every $x \in \tilde{M}$. Notice that such a sequence remains exact when it is tensored with $\mathcal{F}$.

Consider the natural map
(1) $\Gamma_\Phi(M, \Delta_\bullet \otimes \mathcal{F}) \otimes \Gamma_\Psi(M, \mathcal{G}) \to \Gamma_{\Phi\cap\Psi}(M, \Delta_\bullet \otimes \mathcal{F} \otimes \mathcal{G})$.
It is a chain map and hence induces, for any m, a map

$$H^\Phi_m(M, \mathcal{F}) \otimes H^0_\Psi(M, \mathcal{G}) \xrightarrow{\cap} H^{\Phi\cap\Psi}_m(M, \mathcal{F} \otimes \mathcal{G})$$

denoted by $\cap$. If $0 \to \mathcal{F}' \to \mathcal{F} \to \mathcal{F}'' \to 0$ is an exact sequence which remains exact when tensored with $\mathcal{G}$, then the fact that (1) is a chain map implies that

$$\partial(\alpha \cap s) = \partial\alpha \cap s$$

for $\alpha \in H^\Phi_m(M, \mathcal{F}'')$ and $s \in H^0_\Psi(M, \mathcal{G})$. Here ∂ is the connecting homomorphism in the long exact sequence associated to the exact sequence $0 \to \mathcal{F}' \to \mathcal{F} \to \mathcal{F}'' \to 0$ (resp. $0 \to \mathcal{F}' \otimes \mathcal{G} \to \mathcal{F} \otimes \mathcal{G} \to \mathcal{F}'' \otimes \mathcal{G} \to 0$), cf. Prop. 5.9.

Let m and $\alpha \in H^\Phi_m(M, \mathcal{F})$ be fixed. We conclude from Remark 5.10 that the functors $F^q(-)$,

$$F^q(\mathcal{G}) = H^{\Phi\cap\Psi}_{m-q}(M, \mathcal{F} \otimes \mathcal{G}), q \in \mathbf{Z},$$

with connecting homomorphism $(-1)^m$ times that in $H^{\Phi\cap\Psi}_*(M, -)$ (cf. Prop. 5.9) is an (exact) $\mathfrak{E}$-connected sequence of functors.
Hence the natural transformation

$$\begin{aligned} H^0_\Psi(M, \mathcal{G}) &\longrightarrow F^0(\mathcal{G}) = H^{\Phi\cap\Psi}_m(M, \mathcal{F} \otimes \mathcal{G}) \\ \beta &\longmapsto \alpha \cap \beta \end{aligned}$$

has a unique extension

$$\begin{aligned} \alpha\cap : H^p_\Psi(M, \mathcal{G}) &\longrightarrow H^{\Phi\cap\Psi}_{m-p}(M, \mathcal{F} \otimes \mathcal{G}) \quad (p \geq 0) \\ \beta &\longmapsto \alpha \cap \beta \end{aligned}$$

to a morphism of $\mathfrak{E}$-connected sequences of functors ([B, II.6.2]). The uniqueness of $\alpha\cap$ implies that $\alpha \cap \beta$ is linear in α. Of course it is linear in β. Thus we obtain a bilinear *cap product*

$$H^\Phi_m(M, \mathcal{F}) \otimes H^p_\Psi(M, \mathcal{G}) \xrightarrow{\cap} H^{\Phi\cap\Psi}_{m-p}(M, \mathcal{F} \otimes \mathcal{G}).$$

Proposition 8.1. i) Let $0 \to \mathcal{G}' \to \mathcal{G} \to \mathcal{G}'' \to 0$ be an element of $\mathfrak{E}$ and $\alpha \in H^\Phi_m(M, \mathcal{F}), \beta \in H^p_\Psi(M, \mathcal{G}'')$. Then

$$\partial(\alpha \cap \beta) = (-1)^m \alpha \cap \delta\beta \in H^{\Phi\cap\Psi}_{m-p-1}(M, \mathcal{G}')$$

(Here δ is the connecting homomorphism of $H^p_\Psi(M,-)$).
ii) Let $0 \to \mathcal{F}' \to \mathcal{F} \to \mathcal{F}'' \to 0$ be an element of $\mathfrak{E}$ and $\alpha \in H^\Phi_m(M,\mathcal{F}''), \beta \in H^p_\Psi(M,\mathcal{G})$. Then

$$\partial(\alpha \cap \beta) = \partial\alpha \cap \beta.$$

iii) Let $\mathcal{H}$ be another sheaf on M and Θ be another paracompactifying family of supports. Let $\alpha \in H^\Phi_m(M,\mathcal{F}), \beta \in H^p_\Psi(M,\mathcal{G}), \gamma \in H^q_\Theta(M,\mathcal{H})$. Then

$$(\alpha \cap \beta) \cap \gamma = \alpha \cap (\beta \cup \gamma) \in H^{\Phi\cap\Psi\cap\Theta}_{m-p-q}(M, \mathcal{F} \otimes \mathcal{G} \otimes \mathcal{H}).[1]$$

These statements may be proved by the same arguments as those given in [B, p. 212].

If M is a *paracompact* space, then we may define a cap product for arbitrary families Φ, Ψ of supports provided the coefficient sheaves $\mathcal{F}, \mathcal{G}$ are locally constant. This is done as follows. Let $A \in \Phi$ and $B \in \Psi$. By Prop. 7.9 there are *paracompactifying* families $\Phi(A), \Phi(B), \Phi(A \cap B)$ containing $A, B, A \cap B$ respectively such that $\Phi(A) \cap \Phi(B) = \Phi(A \cap B)$ and $H^A_*(M,\mathcal{F}) \xrightarrow{\sim} H^{\Phi(A)}_*(M,\mathcal{F}), H^*_B(M,\mathcal{G}) \xrightarrow{\sim} H^*_{\Phi(B)}(M,\mathcal{G})$ and $H^{A\cap B}_*(M,\mathcal{F}\otimes\mathcal{G}) \xrightarrow{\sim} H^{\Phi(A\cap B)}_*(M,\mathcal{F}\otimes\mathcal{G})$. Now we define a cap product

(*) $\quad H^A_m(M,\mathcal{F}) \otimes H^p_B(M,\mathcal{G}) \xrightarrow{\cap} H^{A\cap B}_{m-p}(M,\mathcal{F}\otimes\mathcal{G})$
by the commutative diagram

$$\begin{array}{ccc} H^A_m(M,\mathcal{F}) \otimes H^p_B(M,\mathcal{G}) & \xrightarrow{\cap} & H^{A\cap B}_{m-p}(M,\mathcal{F}\otimes\mathcal{G}) \\ \cong\downarrow & & \downarrow\cong \\ H^{\Phi(A)}_m(M,\mathcal{F}) \otimes H^p_{\Phi(B)}(M,\mathcal{G}) & \xrightarrow{\cap} & H^{\Phi(A\cap B)}_{m-p}(M,\mathcal{F}\otimes\mathcal{G}). \end{array}$$

This definition does not depend on the choice of the families $\Phi(-)$ (easy exercise, use Remark 7.10). If $A' \in \Phi, B' \in \Psi$ and $A \subset A', B \subset B'$, then we may choose the families $\Phi(-)$ in such a way that $\Phi(A) \subset \Phi(A'), \Phi(B) \subset \Phi(B')$ and $\Phi(A \cap B) \subset \Phi(A' \cap B')$ (Remark 7.10). Then it is easily seen that the diagram

$$\begin{array}{ccc} H^A_m(M,\mathcal{F}) \otimes H^p_B(M,\mathcal{G}) & \longrightarrow & H^{A\cap B}_{m-p}(M,\mathcal{F}\otimes\mathcal{G}) \\ \downarrow & & \downarrow \\ H^{A'}_m(M,\mathcal{F}) \otimes H^p_{B'}(M,\mathcal{G}) & \longrightarrow & H^{A'\cap B'}_{m-p}(M,\mathcal{F}\otimes\mathcal{G}) \end{array}$$

commutes. Thus the cap products (*) fit together to give a cap product

[1] For the definition of the cup product $\cup$ see [B, II.7.5].

$$H_M^{\Phi}(M,\mathcal{F})\otimes H_{\Psi}^{p}(M,\mathcal{G})\xrightarrow{\cap} H_{m-p}^{\Phi\cap\Psi}(M,\mathcal{F}\otimes\mathcal{G}).$$

If Φ and Ψ are paracompactifying, then the cap product defined in this way is the same as the one we defined before. This is clear since we can choose the families $\Phi(A)$ and $\Phi(B)$ as subfamilies of Φ and Ψ. Obviously the statements of Prop. 8.1 remain true for the cap product defined in this way for arbitrary support families and locally constant coefficient sheaves.

Now we want to derive an important formula for the cap product.

Let $f: M \to N$ be a map between locally complete spaces and $\mathcal{F},\mathcal{G}$ be sheaves on N. Let Φ and Φ' (resp. Ψ and Ψ') be families of supports on M (resp. N). Suppose that f is proper and semialgebraic on Φ. Furthermore we assume that $f(\Phi)\subset\Psi, f(\Phi\cap\Phi')\subset\Psi\cap\Psi'$ and $f^{-1}(\Psi')\subset\Phi'$. Then f induces homomorphisms
$f_*: H_*^{\Phi}(M,f^*\mathcal{F}) \to H_*^{\Psi}(N,\mathcal{F}), f_*: H_*^{\Phi\cap\Phi'}(M,f^*(\mathcal{F}\otimes\mathcal{G})) \to H_*^{\Psi\cap\Psi'}(N,\mathcal{F}\otimes\mathcal{G})$, $f^*: H_{\Psi'}^*(N,\mathcal{G}) \to H_{\Phi'}^*(M,f^*\mathcal{G})$ (cf. §6 and II, §2). Assume that at least one of the following two conditions is satisfied:
a) Φ,Φ',Ψ,Ψ' are paracompactifying.
b) M and N are paracompact and $\mathcal{F}$ and $\mathcal{G}$ are locally constant.
Then we have

Proposition 8.2. $f_*(\alpha\cap f^*\beta) = f_*(\alpha)\cap\beta \in H_{m-p}^{\Psi\cap\Psi'}(N,\mathcal{F}\otimes\mathcal{G})$ for $\alpha\in H_m^{\Phi}(M,f^*\mathcal{F})$ and $\beta\in H_{\Psi}^{p}(N,\mathcal{G})$.

Proof. Suppose we are in case b). Obviously it suffices to consider the case where $\Phi = cld(A), \Phi' = cld(A'), A, A' \in \bar{\mathcal{T}}(M)$ and $\Psi = cld(B), \Psi' = cld(B'), B, B' \in \bar{\mathcal{T}}(N)$.

We choose paracompactifying families of supports $\Phi(A)$ on $M, \Psi(B), \Psi(B')$ and $\Psi(B\cap B')$ on N such that

i) $\Psi(B)\cap\Psi(B') = \Psi(B\cap B')$
ii) $A\in\Phi(A), B\in\Psi(B), B'\in\Psi(B'), B\cap B'\in\Psi(B\cap B')$
iii) $H_*^{A}(M,f^*\mathcal{F}) = H_*^{\Phi(A)}(M,f^*\mathcal{F}), H_*^{B}(N,\mathcal{F}) = H_*^{\Psi(B)}(N,\mathcal{F})$,
$H_*^{B\cap B'}(N,\mathcal{F}\otimes\mathcal{G}) = H_*^{\Psi(B\cap B')}(N,\mathcal{F}\otimes\mathcal{G}), H_{B'}^*(N,\mathcal{G}) = H_{\Psi(B')}^*(N,\mathcal{G})$
iv) f is proper on $\Phi(A)$ and $f(\Phi(A))\subset\Psi(B)$.

This is possible by Prop. 7.9: First we choose $\Psi(B),\Psi(B'),\Psi(B\cap B')$. Let $B_1\in\Psi(B)$ be a neighbourhood of B and A_1 be a closed locally semialgebraic neighbourhood of A in M such that $f \mid A_1 : A_1 \to N$ is proper. A_1 exists by Prop. II.8.2. Now we may choose $\Phi(A)$ as a family of subsets of $A_1\cap f^{-1}(B_1)$ (cf. Prop. 7.9). Then iv) is satisfied.

We may replace A by $\Phi(A)$ and B by $\Psi(B)$. We may also replace B' by $\Psi(B')$ and A' by the (paracompactifying) family $f^{-1}(\Psi(B'))$. Now we are in case a).
Hence we assume from now on that all support families are paracompactifying. The diagram

$$\begin{array}{ccc} f_!(\Delta_\bullet(M)\otimes f^*\mathcal{F})\otimes\mathcal{G} & \longrightarrow & f_!(\Delta_\bullet(M)\otimes f^*\mathcal{F}\otimes f^*\mathcal{G}) \\ \downarrow & & \downarrow \\ (\Delta_\bullet(N)\otimes\mathcal{F})\otimes\mathcal{G} & \xrightarrow{\text{id}} & \Delta_\bullet(N)\otimes\mathcal{F}\otimes\mathcal{G} \end{array}$$

commutes. Here the vertical arrows are the maps induced by the map $\alpha: f_!\Delta_\bullet(M) \to \Delta_\bullet(N)$ defined in §6. (They induce the homomorphisms in homology, cf. §6). The upper horizontal arrow is the obvious map. It is an isomorphism by Lemma 6.1. This diagram yields the commutative diagram

$$\begin{array}{ccc}
\Gamma_\Phi(M, \Delta_\bullet(M) \otimes f^*\mathcal{F}) \otimes \Gamma_{\Phi'}(M, f^*\mathcal{G}) & \longrightarrow & \Gamma_{\Phi\cap\Phi'}(M, \Delta_\bullet(M) \otimes f^*(\mathcal{F} \otimes \mathcal{G})) \\
\text{id} \otimes f^* \uparrow & & \\
\Gamma_\Phi(M, \Delta_\bullet(M) \otimes f^*\mathcal{F}) \otimes \Gamma_{\Psi'}(N, \mathcal{G}) & & \downarrow \\
\downarrow & & \\
\Gamma_\Psi(N, f_!(\Delta_\bullet(M) \otimes f^*\mathcal{F})) \otimes \Gamma_{\Psi'}(N, \mathcal{G}) & \longrightarrow & \Gamma_{\Psi\cap\Psi'}(N, f_!(\Delta_\bullet(M) \otimes f^*(\mathcal{F} \otimes \mathcal{G}))) \\
\downarrow & & \downarrow \\
\Gamma_\Psi(N, \Delta_\bullet(N) \otimes \mathcal{F}) \otimes \Gamma_{\Psi'}(N, \mathcal{G}) & \longrightarrow & \Gamma_{\Psi\cap\psi'}(N, \Delta_\bullet(N) \otimes \mathcal{F} \otimes \mathcal{G})
\end{array}$$

The commutativity of this diagram implies that the following diagram commutes for $j = 0$:

$$\begin{array}{ccc}
H_i^\Phi(M, f^*\mathcal{F}) \otimes H^j_{\Phi'}(M, f^*\mathcal{G}) & \xrightarrow{\cap} & H_{i-j}^{\Phi\cap\Phi'}(M, f^*(\mathcal{F} \otimes \mathcal{G})) \\
\text{id} \otimes f^* \uparrow & & \\
H_i^\Phi(M, f^*\mathcal{F}) \otimes H^j_{\Psi'}(N, \mathcal{G}) & & \downarrow f_* \\
f_* \otimes \text{id} \downarrow & & \\
H_i^\Psi(N, \mathcal{F}) \otimes H^j_{\Psi'}(N, \mathcal{G}) & \xrightarrow{\cap} & H_{i-j}^{\Psi\cap\Psi'}(N, \mathcal{F} \otimes \mathcal{G}).
\end{array}$$

If $\alpha \in H_i^\Phi(M, f^*\mathcal{F})$, then both families of maps

$$\begin{aligned}
H^j_{\Psi'}(N, \mathcal{G}) &\rightrightarrows \quad H_{i-j}^{\Psi\cap\Psi'}(N, \mathcal{F} \otimes \mathcal{G}) \quad (j \geq 0) \\
\beta &\longmapsto f_*(\alpha \cap f^*\beta), \\
\beta &\longmapsto f_*(\alpha) \cap \beta,
\end{aligned}$$

are morphisms of $\mathfrak{C}$-connected sequences of functors. We just saw that they coincide for $j = 0$. Since $H^j_{\Psi'}(N, -)$ is universal, we conclude that they coincide for every j [B, II.6.2]. The proof of Prop. 8.2 is finished.

§9 - Poincaré duality

By use of spectral sequences it is an easy matter to derive the Poincaré duality. Our method is an adaption of Bredon's method ([B, V.9]) to the semialgebraic setting.
Let M be a locally complete space over R of finite dimension n. Let $\mathcal{F}$ be a sheaf and Φ be a family of supports on M. We consider the cochain complex of sheaves
$0 \to \Delta^{-n} \otimes \mathcal{F} \to \Delta^{-n+1} \otimes \mathcal{F} \to \ldots \to \Delta^0 \otimes \mathcal{F} \to 0$.
(Recall from §5 $\Delta^k := \Delta_{-k} = \Delta_{-k}(M, \Lambda)$). Let

$$0 \to \Delta^q \otimes \mathcal{F} \to \mathcal{C}^0(\Delta^q \otimes \mathcal{F}) \to \mathcal{C}^1(\Delta^q \otimes \mathcal{F}) \to \ldots$$

be the canonical flabby Godement resolution of $\Delta^q \otimes \mathcal{F} (q \in \mathbb{Z})$ and let $\mathcal{C}^{pq} := \mathcal{C}^p(\Delta^q \otimes \mathcal{F})$. Applying $\Gamma_\Phi(M, -)$ we obtain a double complex of Λ-modules $(\Gamma_\Phi(M, \mathcal{C}^{pq}) \mid p, q \in \mathbb{Z})$. It is bounded below and to the left. Hence it yields two convergent spectral sequences

$$'E_2^{pq} = H_\Phi^p(M, \mathcal{H}^q(\Delta^\bullet \otimes \mathcal{F})) \Longrightarrow {}'E^{p+q} = H^{p+q}(C^\bullet)$$
$$''E_2^{pq} = H^p(H_\Phi^q(M, \Delta^\bullet \otimes \mathcal{F})) \Longrightarrow {}''E^{p+q} = H^{p+q}(C^\bullet)$$

where $C^\bullet$ is the total complex which is associated to the double complex (cf. [B, IV.1]). Observe that $\mathcal{H}^q(\Delta^\bullet \otimes \mathcal{F})$ is just the homology sheaf $\mathcal{H}_{-q}(M, \mathcal{F})$ introduced in §5.
Now we assume that at least one of the following two conditions is satisfied:
a) Φ is paracompactifying
b) M is of type (L) and $\mathcal{F}$ is locally constant.
Then $\Delta^q \otimes \mathcal{F}$ is Φ-acyclic (cf. Remark 5.4) and the second sequence splits. The edge homomorphism $''E^{p0} = H^p(\Gamma_\Phi(\Delta^\bullet \otimes \mathcal{F})) \to H^p(C^\bullet)$ is an isomorphism. Note that $H^p(\Gamma_\Phi(\Delta^\bullet \otimes \mathcal{F})) = H_{-p}^\Phi(M, \mathcal{F})$. Therefore we obtain

Theorem 9.1. There is a spectral sequence

$$H_\Phi^p(M, \mathcal{H}_q(M, \mathcal{F})) \Longrightarrow H_{q-p}^\Phi(M, \mathcal{F}).$$

Now we assume that M is a Λ-**homology manifold**. This means by definition (Def. 1 in §3) that $\mathcal{H}_k(M, \Lambda) = 0$ for $k \neq n$ and $\mathcal{H}_n(M, \Lambda)$ is locally constant with stalks isomorphic to Λ. As usual we simply write $\mathcal{H}_k$ instead of $\mathcal{H}_k(M, \Lambda)$. The „sheaf of orientation" $\mathcal{H}_n$ is also denoted by $\mathcal{O}$.

Lemma 9.2. We have

$$\mathcal{H}_k(M, \mathcal{G}) = \begin{cases} 0 & \text{if } k \neq n \\ \mathcal{O} \otimes \mathcal{G} & \text{if } k = n \end{cases}$$

for every sheaf $\mathcal{G}$ on M.

Proof. We claim that the canonical map $\mathcal{H}_k \otimes \mathcal{G} \to \mathcal{H}_k(M, \mathcal{G})$ is an isomorphism. Consider the map which is induced in the stalks in some point $x \in \tilde{M}$. It is just the map $H_k((\Delta_\bullet)_x) \otimes \mathcal{G}_x \to H_k((\Delta_\bullet)_x \otimes \mathcal{G}_x)$. Since $(\Delta_\bullet)_x$ is torsion free, the universal coefficient formula ([Sp, 5.2.14]) shows that this map is indeed an isomorphism.

From Theorem 9.1 and Lemma 9.2 we get the *Poincaré-duality* theorem.

Theorem 9.3. Let M be a Λ-homology-manifold of dimension n. Assume either that Φ is paracompactifying or that M is of type (L) and $\mathcal{F}$ is locally constant. Then there is a canonical isomorphism

$$D: H^q_\Phi(M, \mathcal{O} \otimes \mathcal{F}) \to H^\Phi_{n-q}(M, \mathcal{F})$$

for every $q \geq 0$.

We consider the situation of Theorem 9.3 and assume that M is Λ-orientable (§3, Def. 2). We choose a Λ-orientation (i.e. an isomorphism)

$$\omega : \Lambda_M \xrightarrow{\sim} \mathcal{O}$$

and denote the fundamental class of M with respect to this orientation by [M]. Recall that $[M] = \omega(\bar{1}) \in \Gamma(M, \mathcal{H}_n) = H_n(M, \Lambda)$ where $\bar{1}$ denotes the element $(\ldots, 1, \ldots) \in \prod_{\pi_0(M)} \Lambda = \Gamma(M, \Lambda_M) = H^0(M, \Lambda_M)$ (cf. Def. 3 in §3). It is easily seen by a look at the double complex which yields the spectral sequence 9.1 that the duality isomorphism $D: H^0(M, \mathcal{O}) = \Gamma(M, \mathcal{H}_n) \to H_n(M, \Lambda)$ is the canonical isomorphism between $\Gamma(M, \mathcal{H}_n)$ and $H_n(M, \Lambda)$ (cf. Prop. 5.12). Since we always identify $\Gamma(M, \mathcal{H}_n)$ with $H_n(M, \Lambda)$, we regard $D: H^0(M, \mathcal{O}) \to H_n(M, \Lambda)$ as the identity map.

Now we consider the duality isomorphism

$$\tilde{D}: H^q_\Phi(M, \mathcal{F}) \xrightarrow[\sim]{\omega_*} H^q_\Phi(M, \mathcal{O} \otimes \mathcal{F}) \xrightarrow[\sim]{D} H^\Phi_{n-q}(M, \mathcal{F}).$$

Suppose that either
a) Φ is paracompactifying
or
b) M is paracompact and $\mathcal{F}$ is locally constant.
Then the cap product $H_n(M, \Lambda) \otimes H^q_\Phi(M, \mathcal{F}) \to H^\Phi_{n-q}(M, \mathcal{F})$ is defined (cf. §8). We have

Proposition 9.4. $\tilde{D}$ is the cap product with the fundamental class $[M]$, i.e.

$$\tilde{D}(\alpha) - [M] \cap \alpha$$

for every $\alpha \in H^q_\Phi(M, \mathcal{F})$.

Proof. If b) is satisfied, it suffices to consider the case $\Phi = cld(A), A \in \bar{\mathcal{T}}(M)$. Choosing a paracompactifying family $\Phi(A)$ with $H^*_A(M, \mathcal{F}) = H^*_{\Phi(A)}(M, \mathcal{F})$ and $H^A_*(M, \mathcal{F}) = H^{\Phi(A)}_*(M, \mathcal{F})$ (cf. §8) we see that case b) follows from case a).
So assume from now on that Φ is paracompactifying. The family of maps $\tilde{D}: H^q_\Phi(M, \mathcal{F}) \to H^\Phi_{n-q}(M, \mathcal{F}), q \in \mathbf{Z}$, is a morphism of connected sequences of functors (cf. [B, IV.1]). For $q = 0$ it coincides with the morphism of connected sequences of functors

$$\begin{aligned} [M]\cap : H^q_\Phi(M, \mathcal{F}) &\longrightarrow H^\Phi_{n-q}(M, \mathcal{F}) \\ \alpha &\longmapsto [M] \cap \alpha. \end{aligned}$$

This can be easily seen by looking at the double complex which yields the spectral sequences. Now we conclude from [B, II.6.2] that the maps $\tilde{D}$ and $[M]\cap$ are identical for every q.

Let Ψ be another family of supports and $\mathcal{G}$ be another sheaf on M. In case a) assume that Ψ is paracompactifying and in case b) that $\mathcal{G}$ is locally constant.

Corollary 9.5. Let $\alpha \in H^p_\Phi(M,\mathcal{F})$ and $\beta \in H^q_\Psi(M,\mathcal{G})$. Then

$$\tilde{D}(\alpha \cup \beta) = \tilde{D}(\alpha) \cap \beta.$$

Proof. $[M] \cap (\alpha \cup \beta) = ([M] \cap \alpha) \cap \beta$ by Prop. 8.1.

§10 - Extension of the base field

Let $S \supset R$ be a real closed field extension of R and M be a locally complete space over R. We use the notation from II, §6. So we denote the space obtained from M by extension of the base field by $M(S)$ and the natural projection $\widetilde{M(S)} \rightarrow \tilde{M}$ by π. If $\mathcal{G}$ is a sheaf on M, then $\mathcal{G}(S)$ is the extension of $\mathcal{G}$ defined by $\pi^*\tilde{\mathcal{G}} = \widetilde{\mathcal{G}(S)}$. We usually omit the groundring Λ in our notation.

Let $U \in \mathring{T}(M)$ be a paracompact open locally semialgebraic subset and $\varphi: X \xrightarrow{\sim} U$ be a triangulation. By Tarski's principle the extension $\varphi_S: X(S) \xrightarrow{\sim} U(S)$ is a triangulation of $U(S)$. We have a canonical map

$$C_\bullet(U,\varphi) \xrightarrow{\varepsilon} C_\bullet(U(S),\varphi_S).$$

Namely, both chain complexes coincide and ε is simply the identity map. Now it is easy to see that these maps ε induce a chain map of sheaves

$$\varepsilon: \Delta_\bullet(M)(S) \longrightarrow \Delta_\bullet(M(S)).$$

Proposition 10.1. ε is a quasiisomorphism

Proof. We may assume that M is semialgebraic. We choose a triangulation $\Phi: X \xrightarrow{\sim} M$ and identify M with X by φ. Let σ be an open simplex of X and $U := \mathrm{St}_X(\sigma)$. Let $y \in \widetilde{\sigma(S)}$ and $x = \pi(y) \in \tilde{\sigma}$. Then x (resp. y) has a fundamental system of neighbourhoods consisting of sets $\widetilde{\mathrm{St}_\psi(z)}$ (resp. $\widetilde{\mathrm{St}_\chi(w)}$) where Ψ (resp. χ) is a triangulation of X (resp. $X(S)$) refining the given triangulation and z (resp. w) is a point in σ (resp. $\sigma(S)$). Since $U = \mathrm{St}_X(z)$ and $U(S) = \mathrm{St}_{X(S)}(\sigma(S)) = \mathrm{St}_{X(S)}(w)$ for $z \in \sigma$ and $w \in \sigma(S)$, we conclude from Proposition 3.1 that the horizontal arrows in the commutative diagram

$$\begin{array}{ccc} H_k(U) & \xrightarrow{\cong} & \mathcal{H}_k(M)_x = \mathcal{H}_k(\Delta_\bullet(M)(S))_y \\ \cong\downarrow & & \downarrow "\varepsilon" \\ H_k(U(S)) & \xrightarrow{\cong} & \mathcal{H}_k(M(S))_y \end{array}$$

are isomorphisms. The left vertical arrow is just the canonical isomorphism $H_k(U) \xrightarrow{\sim} H_k(U(S))$ from Theorem 2.13. We see that ε induces quasiisomorphisms in the stalks and hence is a quasiisomorphism.

Now let $\mathcal{F}$ be a sheaf and Φ be a family of supports on M. Recall that $\Phi(S)$ is the support family on $M(S)$ generated by the sets $A(S), A \in \Phi$. It is paracompactifying if Φ is paracompactifying. We assume that at least one of the following two conditions is satisfied:
a) M is of type (L), $\dim M < \infty$ and $\mathcal{F}$ is locally constant.
b) Φ is paracompactifying and $\dim \Phi < \infty$.
Then we know from Prop. 5.8 that $H_k^\Phi(M,\mathcal{F}) = \mathrm{H}_\Phi^{-k}(M, \Delta_\bullet(M) \otimes \mathcal{F})$ and $H_k^{\Phi(S)}(M(S),\mathcal{F}(S)) = \mathrm{H}_{\Phi(S)}^{-k}(M(S), \Delta^\bullet(M(S)) \otimes \mathcal{F}(S))$.
Since the sheaves Δ^k are torsion free and ε is a quasiisomorphism (Prop. 10.1), the canonical map $(\Delta^\bullet(M) \otimes \mathcal{F})(S) \rightarrow \Delta^\bullet(M(S)) \otimes \mathcal{F}(S)$ is also a quasiisomorphism. Thus we have

$$H_k^{\Phi(S)}(M(S),\mathcal{F}(S)) = \mathsf{H}_{\Phi(S)}^{-k}(M(S),(\Delta^\bullet(M)\otimes\mathcal{F})(S)).$$

We know from Theorem II.10.4 that the canonical homomorphism

$$\mathsf{H}_\Phi^*(M,\mathcal{G}^\bullet) \xrightarrow{\pi^*} \mathsf{H}_{\Phi(S)}^*(M(S),\mathcal{G}^\bullet(S))$$

is an isomorphism for every complex $\mathcal{G}^\cdot$ of sheaves. Altogether we obtain

Theorem 10.2. The canonical map

$$\pi^* : H_*^\Phi(M,\mathcal{F}) \to H_*^{\Phi(S)}(M(S),\mathcal{F}(S))$$

is an isomorphism.

Now we assume that $\dim M < \infty$. Recall that $c(M)(S)$ { resp. $pc(M)(S)$} is in general a proper subfamily of $c(M(S))$ { resp. $pc(M(S))$}. Nevertheless we obtain by Theorem II.10.5

Theorem 10.3. Let $\mathcal{F}$ be any sheaf on M.

a) The canonical map

$$\pi^* : H_*^c(M,\mathcal{F}) \to H_*^c(M(S),\mathcal{F}(S))$$

is an isomorphism.

b) If M is paracompact, the canonical map

$$\pi^* : H_*^{pc}(M,\mathcal{F}) \to H_*^{pc}(M(S),\mathcal{F}(S))$$

is an isomorphism.

§11 - Comparison with topological Borel-Moore-homology

Let M be a taut locally complete space over R and $X = \tilde{M}$ be the associated abstract space. Since X is regular and taut (I, §3), the space $X^{\max}$ of closed points is a locally compact topological space (I.4.5). Hence the topological Borel-Moore-homology groups ${}^{BM}H_*^{\Phi}(X^{\max}, \mathcal{F})$ of $X^{\max}$ are defined (for every sheaf $\mathcal{F}$ and every support family Φ on $X^{\max}$, cf. [BM], [B, Chap. V]). We may ask the question whether there is a relation between these groups and the semialgebraic Borel-Moore-homology of M.

Example 11.1. If M is a partially complete space over $\mathbb{R}$, then $X^{\max}$ is just the topological space M_{top} (i.e. M equipped with its strong topology).

If $\mathcal{F}$ is a sheaf on M, then $\mathcal{F} \mid X^{\max}$ is (as usual) the restriction of $\tilde{\mathcal{F}}$ to $X^{\max}$. For every family Φ of supports on M we denote, as in chap. II, the family $\tilde{\Phi} \cap X^{\max}$ by Φ^m. If Y is a topological space (e.g. $Y = X^{\max}$ or $Y = M_{\text{top}}$), then we denote the family of compact subsets of Y by $c_{\text{top}}(Y)$ or simply by c_{top}.

Example 11.2. It is easily seen that $\bar{\gamma}(M)^m = \bar{\gamma}(X) \cap X^{\max}$ is the family $c_{\text{top}}(X^{\max})$.

Proposition 11.3. Suppose that M is partially complete. Then

$$U \mapsto \Gamma_{c_{\text{top}}(U)}(U, \Delta_\bullet(M, \Lambda) \mid X^{\max}) \quad (U \subset X^{\max} \text{ open})$$

is a flabby coresolution of Λ on $X^{\max}$ (cf. [B, V.11.4] for the definition of coresolutions).

Proof. First we observe that $\Gamma_{c_{\text{top}}}(-, \Delta_k \mid X^{\max})$ is a flabby cosheaf ([B, V.1]) for every k since $\Delta_k \mid X^{\max}$ is a c_{top}-soft sheaf by (II.4.8). Let $x \in X^{\max}$. It suffices to prove that x has a fundamental system of open neighbourhoods U such that $H_k(\Gamma_{c_{\text{top}}}(U, \Delta_\bullet \mid X^{\max}))$ is 0 for $k \neq 0$ and isomorphic to Λ for $k = 0$. Now x posesses a fundamental system of neighbourhoods $\tilde{V} \cap X^{\max}$ with $V \subset M$ open and semialgebraic and contractible. Consider such a set V and let $U = \tilde{V} \cap X^{\max}$. Then $\Gamma_{c_{\text{top}}(U)}(U, \Delta_\bullet \mid X^{\max}) = \Gamma_{c_{\text{top}}(X^{\max})|U}(X^{\max}, \Delta_\bullet \mid X^{\max})$. By Example 11.2 we have $c_{\text{top}}(X^{\max}) = \bar{\gamma}(M) \cap X^{\max}$. Since M is partially complete, $\bar{\gamma}(M)$ is the family c of complete semialgebraic subsets of M.
Hence $c_{\text{top}}(X^{\max}) \mid U = (c \mid V)\cap X^{\max}$. Now we conclude from (II.3.5) that $\Gamma_{c_{\text{top}}|U}(X^{\max}, \Delta_\bullet \mid X^{\max}) = \Gamma_{c|V}(M, \Delta_\bullet) = \Gamma_c(V, \Delta_\bullet)$. Since V is contractible, we see that $H_k(\Gamma_{c_{\text{top}}}(U, \Delta_\bullet \mid X^{\max})) = H_k^c(V, \Lambda) \cong \begin{cases} 0 & k \neq 0 \\ \Lambda & k = 0 \end{cases}$. Hence $\Gamma_{c_{\text{top}}}(-, \Delta_\bullet \mid X^{\max})$ is indeed a coresolution of Λ on $X^{\max}$.

From now on we assume that $\dim M < \infty$. Then $\dim_\Lambda X^{\max} < \infty$ by Lemma II.9.8. Since $X^{\max}$ is also clc_Λ^∞ by Lemma II.9.9 the hypotheses of Theorem V.11.15 of [B] are satisfied and we conclude by use of this theorem and Prop. 11.3 that the following holds.

Corollary 11.4. Suppose M is partially complete. Let Ψ be a family of supports on $X^{\max}$ which is paracompactifying in the classical sense. Then there is a canonical isomorphism

$${}^{BM}H_*^{\Psi}(X^{\max}, \mathcal{G}) \cong H_*(\Gamma_\Psi(X^{\max}, (\Delta_\bullet \mid X^{\max}) \otimes \mathcal{G}))$$

for every sheaf $\mathcal{G}$ on $X^{\max}$.

From Corollary 11.4 we easily derive

Theorem 11.5. Suppose M is partially complete and Φ is a paracompactifying family of supports on M. Then there is a canonical isomorphism

$$H_*^{\Phi}(M,\mathcal{F}) \cong {}^{BM}H_*^{\Phi^m}(X^{\max},\mathcal{F} \mid X^{\max})$$

between the semialgebraic Borel-Moore-homology of M and the topological Borel-Moore-homology of $X^{\max}$ for every sheaf $\mathcal{F}$ on M.

Proof. The support family Φ^m is paracompactifying in the classical sense (Prop. II.1.6). Since $\Gamma_\Phi(M,\mathcal{G}) = \Gamma_{\Phi^m}(X^{\max},\mathcal{G} \mid X^{\max})$ for every sheaf $\mathcal{G}$ on M (Cor. II.3.5), our claim follows from Cor. 11.4.

As a special case (cf. Example 11.1) we obtain

Corollary 11.6. Suppose M is a partially complete space over R. Then there is a canonical isomorphism

$$H_*^{\Phi}(M,\mathcal{F}) \cong {}^{BM}H_*^{\Phi_{\text{top}}}(M_{top},\mathcal{F} \mid M_{\text{top}})$$

for every sheaf $\mathcal{F}$ and every paracompactifying family Φ of supports on M.

Remark 11.7. If $M \mid$ R is not assumed to be taut, then in general $M \subsetneq X^{\max}$ (cf. the example in Remark I.4.6). Nevertheless Prop. 11.3 is true with $X^{\max}$ replaced by M_{top}. Even the hypothesis that M is partially complete is superfluous. One can verify this by the same arguments as in the proof of 11.3 observing that $c_{\text{top}}(M_{\text{top}}) = c(M) \cap M_{\text{top}}$ (cf. [DK, §9]). Hence Corollary 11.4 remains true with $X^{\max}$ replaced by M_{top} (and without the hypothesis that M is partially complete). We see hat the assumption that M is taut is not necessary in Cor. 11.6.

But notice that the hypothesis that M is partially complete is necessary in Cor. 11.6 because only in this case M_{top} is a partially quasicompact subset of X. Here is a counterexample to 11.6 if M is not partially complete.

Example 11.8. Consider the affine semialgebraic space $M :=$ R over R and the sheaf $\mathcal{O}_M$ of semialgebraic functions on M. By semialgebraic and topological Poincaré duality (cf. §10 and [B, V.9]) we conclude that

$$H_1(M,\mathcal{O}_M) \cong H^0(M,\mathcal{O}_M)$$
$${}^{BM}H_1(M_{\text{top}},\mathcal{O}_M \mid M_{top}) \cong H^0(M_{\text{top}},\mathcal{O}_M \mid M_{\text{top}}).$$

But $H^0(M,\mathcal{O}_M)$ is not equal to $H^0(M_{\text{top}},\mathcal{O}_M \mid M_{\text{top}})$ (cf. Example II.5.6.a).

If the coefficient sheaves are assumed to be locally constant, such counterexamples do not exist (cf. Theorem 11.10).

Remark 11.9. Let M be a locally complete (not necessarily taut) space over R. By Cor. 11.4 and Remark 11.7, we have a canonical isomorphism

$$H_*(\Gamma_{\Phi_{\text{top}}}(M_{\text{top}},(\Delta_\bullet \otimes \mathcal{F}) \mid M_{\text{top}})) \xrightarrow{\sim} {}^{BM}H_*^{\Phi_{\text{top}}}(M_{\text{top}},\mathcal{F} \mid M_{\text{top}})$$

for every sheaf $\mathcal{F}$ and every paracompactifying family Φ of supports on M. Hence we have a canonical homomorphism

$$H_*^{\Phi}(M,\mathcal{F}) \to {}^{BM}H_*^{\Phi_{\mathrm{top}}}(M_{\mathrm{top}}, \mathcal{F} \mid M_{\mathrm{top}}).$$

It is compatible with restriction to open locally semialgebraic subsets (cf. [B, V.11]) and, of course, coincides with the canonical isomorphism in Cor. 11.6 if M is partially complete.

Theorem 11.10. Let M be a locally complete (not necessarily taut) space over $\mathbf{R}$ and Φ be a family of supports on M such that each member A of Φ is paracompact. Let $\mathcal{F}$ be a locally constant sheaf on M whose stalks are finitely generated Λ-modules. Then there is a canonical isomorphism

$$H_*^{\Phi}(M,\mathcal{F}) \cong {}^{BM}H_*^{\mathrm{top}}(M_{\mathrm{top}}, \mathcal{F} \mid M_{\mathrm{top}}).$$

Proof. We have

$$H_*^{\Phi}(M,\mathcal{F}) = \varinjlim_{A \in \Phi} H_*(A, \mathcal{F} \mid A)$$

by Prop. 5.7 and

$${}^{BM}H_*^{\Phi_{\mathrm{top}}}(M_{\mathrm{top}}, \mathcal{F} \mid M_{\mathrm{top}}) = \varinjlim_{A \in \Phi} {}^{BM}H_*(A_{\mathrm{top}}, \mathcal{F} \mid A_{\mathrm{top}})$$

by [B, V.5.5]. (Here we need the hypothesis that the stalks of $\mathcal{F}$ are finitely generated). Hence we may assume that $\Phi = cld(M)$ and M is paracompact.

Now we choose a pure completion $M \overset{i}{\hookrightarrow} \bar{M}$ of M ([DK$_3$, II, §9, Def. 2]). Let $\mathcal{G} = i_*\mathcal{F}$ and $A := \bar{M} \setminus M$ (we regard i as an inclusion map). Since $(\bar{M}, M)$ is pure, $\mathcal{G}$ is also locally constant. Then we consider the commutative ladder

$$\begin{array}{ccccccc}
\ldots \to H_n(A, \mathcal{G} \mid A) \longrightarrow & & H_n(\bar{M}, \mathcal{G}) & \xrightarrow{\mathrm{res}} & H_n(M, \mathcal{F}) \to \ldots \\
\downarrow \alpha & & \downarrow \beta & & \downarrow \gamma \\
\ldots \to {}^{BM}H_n(A_{\mathrm{top}}, \mathcal{G} \mid A_{\mathrm{top}}) \to & & {}^{BM}H_n(\bar{M}_{\mathrm{top}}, \mathcal{G} \mid \bar{M}_{\mathrm{top}}) & \xrightarrow{\mathrm{res}} & {}^{BM}H_n(M_{\mathrm{top}}, \mathcal{F} \mid M_{\mathrm{top}}) \to \ldots
\end{array}$$

The horizontal sequences are the long exact sequences of the triples $(\bar{M}, M, A)$ (cf. Prop. 5.11) and $(\bar{M}_{\mathrm{top}}, M_{\mathrm{top}}, A_{\mathrm{top}})$ (cf. [B, Remark after Cor. V.5.9]). The maps α, β, γ are the canonical maps mentioned in Remark 11.9. From Cor. 11.6 we know that α and β are isomorphisms. By the Five Lemma we conclude that γ is also an isomorphism. q.e.d.

Remark 11.11. Let M be a paracompact (locally complete) space over an arbitrary real closed field R and $\mathcal{F}$ be a locally constant sheaf on M, whose stalks are finitely generated Λ-modules. Let $M \subset \bar{M}$ be a pure completion of M and let $X = \tilde{M}, \bar{X} = \tilde{\bar{M}}$. Analogous arguments as those in the proof of 11.10 show that there is a canonical isomorphism

$$H_*^{\Phi}(M,\mathcal{F}) \cong {}^{BM}H_*^{\Phi^m}(\bar{X}^{\max} \cap X, \mathcal{F} \mid \bar{X}^{\max} \cap X)$$

for every family Φ of supports on M. (You have to use the sequence of the triple $(\bar{X}^{\max}, \bar{X}^{\max} \cap X, \bar{X}^{\max} \setminus X)$ instead of the sequence of $(\bar{M}_{\mathrm{top}}, M_{\mathrm{top}}, A_{\mathrm{top}})$ and Theorem 11.5 instead of Cor. 11.6).

§12 - Duals of complexes of sheaves and Borel-Moore-homology

As always we consider a locally complete space M oder R. We will show in this section that the semialgebraic Borel-Moore-homology groups defined in §5 can also be obtained by a definition which is analogous to that originally given by Borel and Moore for locally compact topological spaces ([BM]). Our considerations lead to an important short exact sequence (cf. 12.12 below).

Let $\mathcal{L}$ be a sheaf on M. If $U, V \in \mathring{\mathcal{T}}(M)$ and $V \subset U$, then we have a natural inclusion map $\Gamma_c(V,\mathcal{L}) \subset \Gamma_c(U,\mathcal{L})$ denoted by $i_{U,V}$.

Proposition 12.1. Assume that $\mathcal{L}$ is c-soft. Let $U \in \mathring{\mathcal{T}}(M)$ and $(U_i \mid i \in I)$ be an admissible covering of U by open locally semialgebraic subsets. Then the sequence $\sum\limits_{i,j} \Gamma_c(U_i \cap U_j, \mathcal{L}) \xrightarrow{g} \sum\limits_i \Gamma_c(U_i,\mathcal{L}) \xrightarrow{f} \Gamma_c(U,\mathcal{L}) \to 0$ is exact. Here f is the map $\sum i_{U,U_i}$ and g is the map $\sum\limits_{i,j}(i_{U_i,U_i\cap U_j} - i_{U_j,U_i\cap U_j})$.

Proof. Since each member A of c is semialgebraic, we may assume that U is semialgebraic and that $I = \{1,\ldots,n\}$ for some $n \in \mathbb{N}$.

We want to prove that f is surjective. It suffices to consider the case $n = 2$ (iterate!). Let V_i be an open semialgebraic subset of U_i $(i = 1,2)$ such hat $\bar{V}_i \cap U \subset U_i$ and $V_1 \cup V_2 = U$ (cf. I.5.1). Let $s \in \Gamma_c(U, L)$ and choose some complete semialgebraic subset K of U with $s \mid U \setminus K = 0$. Since $\mathcal{L}$ is c-soft we may extend $s \mid K \cap \bar{V}_1$ to a section $s_1 \in \Gamma_c(U_1,\mathcal{L})$ (II.4.1). Let $s_2 = s - i_{U,U_1}(s_1) \in \Gamma_c(U,\mathcal{L})$. Then $\operatorname{supp}(s_2) \subset K \setminus V_1 \subset V_2 \subset U_2$. Hence $s_2 \in \Gamma_c(U_2,\mathcal{L})$. Obviously we have $f((s_1,s_2)) = s$.
It remains to prove that the kernel of f is equal to the image of g. This is done by induction on n. The case $n = 1$ is trivial. So let $n > 1$ and suppose that $s_i \in \Gamma_c(U_i,\mathcal{L})$ are given $(1 \le i \le n)$ with $f((s_i \mid 1 \le i \le n)) = 0$. Let $W := U_1 \cup \ldots \cup U_{n-1}$ and $s := \sum\limits_{i=1}^{n-1} i_{W,U_i}(s_i) \in \Gamma_c(W,\mathcal{L})$. We have $i_{U,W}(s) = -i_{U,U_n}(s_n)$. Hence $\operatorname{supp}(s) \subset W \cap U_n$ and we may consider s as an element of $\Gamma_c(W \cap U_n,\mathcal{L})$. We already know that f is always surjective. Hence there are sections $t_i \in \Gamma_c(U_i \cap U_n,\mathcal{L}), 1 \le i \le n-1$, such that $s = \sum\limits_{i=1}^{n-1} i_{W\cap U_n, U_i\cap U_n}(t_i)$. Now we define

$$t_{ij} := \begin{cases} t_i & \text{if} \quad i < n \quad \text{and} \quad j = n \\ 0 & \text{else} \end{cases}$$

Let $u_i = s_i - i_{U_i,U_i\cap U_n}(t_i) \in \Gamma_c(U_i,\mathcal{L}), 1 \le i \le n-1$. Then $(s_i \mid 1 \le i \le n) - g((t_{ij} \mid 1 \le i \le n)) = (u_i \mid 1 \le i \le n-1)$. But $f((u_i \mid 1 \le i \le n-1)) = 0$ and hence $(u_i \mid 1 \le i \le n-1)$ lies in the image of g by our induction hypothesis.

Corollary 12.2 Let $\mathcal{L}$ be a c-soft sheaf on M and P be a Λ-module. Then

$$U \mapsto \operatorname{Hom}_\Lambda(\Gamma_c(U,\mathcal{L}), P) \qquad (U \in \mathring{\mathcal{T}}(M))$$

is a sheaf.

This is an immediate consequence of Prop. 12.1 since $\mathrm{Hom}_\Lambda(-,P)$ is a left exact functor. We want to define the dual of a (cochain) complex $\mathcal{L}^\bullet$ of sheaves (cf. [BM], [B, V.1]). First we recall the definition of $\mathrm{Hom}(A^\bullet, B^\bullet)$ for (cochain) complexes $A^\bullet$ and $B^\bullet$ in an abelian category. $\mathrm{Hom}(A^\bullet, B^\bullet)$ is a (cochain) complex with $\mathrm{Hom}(A^\bullet, B^\bullet)^n := \bigoplus_{q-p=n} \mathrm{Hom}(A^p, B^q)$. The differential d is defined as follows: If $f \in \mathrm{Hom}(A^\bullet, B^\bullet)^n$, then $df := d_{B^\bullet} \circ f - (-1)^n \cdot f \circ d_{A^\bullet}$. Now let $\mathcal{L}^\bullet$ be a cochain complex of c-soft sheaves on M. We choose an injective resolution $0 \to \Lambda \to \Lambda^0 \to \Lambda^1 \to 0$ of Λ. This is possible since Λ is a principial ideal domain and hence every Λ-module has an injective resolution of length 2. (N.B. A Λ-module is injective if it is divisible). The *dual* $\mathcal{D}(\mathcal{L}^\bullet)$ of $\mathcal{L}^\bullet$ is defined to be the complex of sheaves

$$U \mapsto \mathrm{Hom}(\Gamma_c(U, \mathcal{L}^\bullet), \Lambda^\bullet) \qquad (U \in \mathring{\mathcal{T}}(M)).$$

Here $\Lambda^\bullet$ denotes the complex $0 \to \Lambda^0 \to \Lambda^1 \to 0$. Note that $\mathcal{D}(\mathcal{L}^\bullet)^n$ is indeed a sheaf (Cor. 12.2). If we chose another injective resolution of Λ, then we would obtain a homotopy equivalent complex of sheaves.

Proposition 12.3. $\mathcal{D}(\mathcal{L}^\bullet)$ is a complex of sa-flabby sheaves. It is torsion free if $\Gamma_c(U, \mathcal{L}^k)$ is divisible for every $U \in \mathring{\mathcal{T}}(M)$ and every k. Moreover, $\Gamma_c(U, \mathcal{D}(\mathcal{L}^\bullet)^n)$ is divisible for every $U \in \mathring{\mathcal{T}}(M)$ and every n if $\mathcal{L}^k$ is torsion free for every k.

This is clear since $\Lambda^\bullet$ consists of injective modules (cf. [B, V.1.11]).

Now we consider the (cochain) complex of sa-flabby (and hence c-soft) sheaves $\Delta^\bullet(M, \Lambda)$. (Recall $\Delta^k(M, \Lambda) = \Delta_{-k}(M, \Lambda)$). We often simply write $\Delta^\bullet$ or $\Delta^\bullet(M)$ instead of $\Delta^\bullet(M, \Lambda)$. Notice that we have a natural map $\Lambda_M \xrightarrow{j} \mathcal{D}(\Delta^\bullet)^0$ which may be described as follows: Let $U \in \mathring{\mathcal{T}}(M)$ be connected and $\Gamma_c(U, \Delta^0) \xrightarrow{\eta} \Lambda$ be the augmentation map. Then j is the map $\Gamma(U, \Lambda_M) = \Lambda = \mathrm{Hom}_\Lambda(\Lambda, \Lambda) \xrightarrow{\circ\eta} \mathrm{Hom}(\Gamma_c(U, \Delta^0), \Lambda) \to \mathrm{Hom}(\Gamma_c(U, \Delta^0), \Lambda^0) \hookrightarrow \Gamma(U, \mathcal{D}(\Delta^\bullet)^0)$.
We also observe that $\mathcal{D}(\Delta^\bullet)^k = 0$ für $k < 0$.

Lemma 12.4. $\Lambda_M \xrightarrow{j} \mathcal{D}(\Delta^\bullet)$ is a sa-flabby resolution of Λ_M.

Proof. From the universal coefficient formula (cf. e.g. [CE, XVII.5.2]) we obtain the exact sequence

$$0 \to \mathrm{Ext}(H^{1-p}(\Gamma_c(U, \Delta^\bullet)), \Lambda) \to H^p(\Gamma(U, \mathcal{D}(\Delta^\bullet))) \to \mathrm{Hom}(H^{-p}(\Gamma_c(U, \Delta^\bullet)), \Lambda) \to 0$$

for every $U \in \mathring{\mathcal{T}}(M)$. Note that $H^{-q}(\Gamma_c(U, \Delta^\bullet)) = H_q^c(U, \Lambda)$. Suppose U is contractible. Then $H_q^c(U, \Lambda)$ is 0 for $q \neq 0$ and isomorphic to Λ for $q = 0$. Hence

$$H^p(\Gamma(U, \mathcal{D}(\Delta^\bullet))) = \begin{cases} \Lambda & \text{for } p = 0 \\ 0 & \text{for } p > 0. \end{cases} \qquad \text{q.e.d.}$$

It is easily seen that there is a natural chain map

$$\mathcal{L}^\bullet \xrightarrow{\alpha} \mathcal{D}(\mathcal{D}(\mathcal{L}^\bullet))$$

for every complex $\mathcal{L}^\bullet$ of c-soft sheaves (cf. [B, V.1.12]). Our next goal is to prove

Proposition 12.5. The natural map

$$\Delta^\bullet(M) \xrightarrow{\alpha} \mathcal{D}(\mathcal{D}(\Delta^\bullet(M)))$$

is a quasiisomorphism.

Prop. 12.5 follows from

Proposition 12.6. Assume that M is semialgebraic. Then the natural map

$$\Gamma(M, \Delta^\bullet) \xrightarrow{\alpha_M} \Gamma(M, \mathcal{D}(\mathcal{D}(\Delta^\bullet)))$$

is a quasiisomorphism.

We need the following well known lemmas.

Lemma 12.7. Let $A^\bullet$ be a complex of free Λ-modules. The chain map

$$\mathrm{Hom}\,(A^\bullet, \Lambda) \to \mathrm{Hom}\,(A^\bullet, \Lambda^\bullet)$$

is a quasiisomorphism.

Lemma 12.8. Let $A^\bullet \to B^\bullet$ be a quasiisomorphism of complexes of Λ-modules. Then the induced map

$$\mathrm{Hom}\,(B^\bullet, \Lambda^\bullet) \to \mathrm{Hom}\,(A^\bullet, \Lambda^\bullet)$$

is a quasiisomorphism.

Proof of Prop. 12.6 if M is complete:
We choose a triangulation $\varphi : X \xrightarrow{\sim} M$. Let $C^k(M, \varphi) := C_{-k}(M, \varphi; \Lambda)$. We know that the canonical map $C^\bullet(M, \varphi) \to \Gamma(M, \Delta^\bullet)$ is a quasiisomorphism (Remark 4.15). Since $C^\bullet(M, \varphi)$ is a free Λ-module of finite rank, we have

$$C^\bullet(M, \varphi) = \mathrm{Hom}\,(\mathrm{Hom}\,(C^\bullet(M, \varphi), \Lambda), \Lambda).$$

Now we consider the commutative diagram

$$\begin{array}{ccc} C^\bullet(M,\varphi) & \longrightarrow & \mathrm{Hom}\,(\mathrm{Hom}\,(C^\bullet(M,\varphi),\Lambda^\bullet),\Lambda^\bullet) \\ \downarrow & & \downarrow \\ \Gamma(M,\Delta^\bullet) & \xrightarrow{\alpha_M} & \mathrm{Hom}\,(\mathrm{Hom}\,(\Gamma(M,\Delta^\bullet),\Lambda^\bullet),\Lambda^\bullet) \\ & & \| \\ & & \Gamma(M,\mathcal{D}(\mathcal{D}(\Delta^\bullet))) \end{array}$$

From the preceding lemmas we know that the upper horizontal and the right vertical arrow are quasiisomorphisms. Hence the lower horizontal arrow is also a quasiisomorphism.
This proves 12.6 if M is complete.

To prove Prop. 12.6 in the general case we choose a completion $i: M \hookrightarrow N$ of M. We regard i as an inclusion map. Let $A := N \setminus M$. Both spaces N and A are complete semialgebraic spaces. Since

$$\Gamma(U \cap A, \Delta^\bullet(A)) = \Gamma_{A\cap U}(U, \Delta^\bullet(N)) \subset \Gamma(U, \Delta^\bullet(N))$$

for $U \in \mathring{\gamma}(N)$ (Prop. 4.12), we have a natural map

$$\mathcal{D}(\Delta^\bullet(N)) \mid A \to \mathcal{D}(\Delta^\bullet(A))$$

of complexes of sheaves on A.

Lemma 12.9. This map induces a quasiisomorphism

$$\Gamma(A, \mathcal{D}(\Delta^\bullet(N)) \mid A) \to \Gamma(A, \mathcal{D}(\Delta^\bullet(A))).$$

Proof. This is clear since both complexes $\mathcal{D}(\Delta^\bullet(N)) \mid A$ and $\mathcal{D}(\Delta^\bullet(A))$ are soft resolutions of Λ_A (by Lemma 12.4) and the map is a morphism of resolutions.

Now Lemma 12.8 implies that the induced map

$$\operatorname{Hom}(\Gamma(A, \mathcal{D}(\Delta^\bullet(A))), \Lambda^\bullet) \to \operatorname{Hom}(\Gamma(A, \mathcal{D}(\Delta^\bullet(N)) \mid A), \Lambda^\bullet)$$

is also a quasiisomorphism. We denote this map by β.
Since $\mathcal{D}(\Delta^\bullet(N))^k$ is *sa*-flabby and hence soft, the natural exact sequence

$$0 \to \Gamma_c(M, \mathcal{D}(\Delta^\bullet(N))) \to \Gamma(N, \mathcal{D}(\Delta^\bullet(N))) \to \Gamma(A, \mathcal{D}(\Delta^\bullet(N)) \mid A) \to 0$$

is exact. It remains exact when we apply the exact functor $\operatorname{Hom}(-, \Lambda^\bullet)$, and it yields the lower row of the following commutative diagram

$$\begin{array}{ccccccc}
0 \to \Gamma(A, \Delta^\bullet(A)) & \longrightarrow & \Gamma(N, \Delta^\bullet(N)) & \xrightarrow{\text{res}} & \Gamma(M, \Delta^\bullet(M)) & \to 0 \\
\alpha_A \downarrow & & & & & \\
\operatorname{Hom}(\Gamma(A, \mathcal{D}(\Delta^\bullet(A))), \Lambda^\bullet) & & \downarrow \alpha_N & & \downarrow \alpha_M & \\
\beta \downarrow & & & & & \\
0 \to \operatorname{Hom}(\Gamma(A, \mathcal{D}(\Delta^\bullet(N)) \mid A), \Lambda^\bullet) & \to & \Gamma(N, \mathcal{D}(\mathcal{D}(\Delta^\bullet(N)))) & \to & \Gamma(M, \mathcal{D}(\mathcal{D}(\Delta^\bullet(M)))) & \to 0.
\end{array}$$

Both rows are exact. We already know that Prop. 12.6 is true in the complete case. Hence α_N and α_A are quasiisomorphisms. Since β is also a quasiisomorphism, the Five Lemma gives Prop. 12.6 in the general case. q.e.d.

Notice that $\Delta^\bullet$ consists of torsion free sheaves and hence $\mathcal{D}(\mathcal{D}(\Delta^\bullet))$ is also torsion free (Prop. 12.3). Thus the quasiisomorphism α induces a quasiisomorphism $\alpha \otimes \mathrm{id} : \Delta^\bullet \otimes \mathcal{F} \to \mathcal{D}(\mathcal{D}(\Delta^\bullet)) \otimes \mathcal{F}$ for every sheaf $\mathcal{F}$ on M.
Now let Φ be a family of supports and assume either that
1) Φ is paracompactifying and $\dim \Phi < \infty$
or that
2) M is of type (L) and $\dim M < \infty$.

Then $\mathrm{H}^*_\Phi(M,\mathcal{G}^\bullet)$ is defined for all complexes $\mathcal{G}^\bullet$ of sheaves (cf. II, §10). The quasiisomorphism $\alpha\otimes\mathrm{id}$ induces an isomorphism in hypercohomology:

$$\mathrm{H}^*_\Phi(M,\Delta^\bullet\otimes\mathcal{F}) \xrightarrow{\sim} \mathrm{H}^*_\Phi(M,\mathcal{D}(\mathcal{D}(\Delta^\bullet))\otimes\mathcal{F}).$$

Recall that $\mathcal{D}(\Delta^\bullet)$ is a sa-flabby resolution of Λ_M. We choose an injective resolution $0\to\Lambda_M\to\mathcal{J}^\bullet$ of Λ_M and a homomorphism of resolutions

$$\mathcal{D}(\Delta^\bullet)\longrightarrow\mathcal{J}^\bullet.$$

Up to homotopy it is uniquely determined. By Lemma 12.8 it induces a quasiisomorphism

$$\mathcal{D}(\mathcal{J}^\bullet)\longrightarrow\mathcal{D}(\mathcal{D}(\Delta^\bullet))$$

since $\mathcal{J}^\bullet$ and $\mathcal{D}(\Delta^\bullet)$ consist of c-soft sheaves and hence $\Gamma_c(U,\mathcal{D}(\Delta^\bullet))\to\Gamma_c(U,\mathcal{J}^\bullet)$ is a quasiisomorphism for $U\in\mathring{T}(M)$. Now $\Gamma_c(U,\mathcal{J}^k)$ is divisible for every $U\in\mathring{T}(M)$ (cf. [B, II.3.3]). Hence $\mathcal{D}(\mathcal{J}^\bullet)$ is torsion free and we conclude that

$$\mathcal{D}(\mathcal{J}^\bullet)\otimes\mathcal{F}\longrightarrow\mathcal{D}(\mathcal{D}(\Delta^\bullet))\otimes\mathcal{F}$$

is a quasiisomorphism. It induces an isomorphism in hypercohomology

$$\mathrm{H}^*_\Phi(M,\mathcal{D}(\mathcal{J}^\bullet)\otimes\mathcal{F}) \xrightarrow{\sim} \mathrm{H}^*_\Phi(M,\mathcal{D}(\mathcal{D}(\Delta^\bullet))\otimes\mathcal{F}).$$

Altogether we obtain a canonical isomorphism

$$\mathrm{H}^*_\Phi(M,\Delta^\bullet\otimes\mathcal{F}) \xrightarrow{\sim} \mathrm{H}^*_\Phi(M,\mathcal{D}(\mathcal{J}^\bullet)\otimes\mathcal{F}).$$

Theorem 12.10. Let Φ be a family of supports on M. Suppose either that Φ is paracompactifying and $\dim\Phi<\infty$ or that M is of type (L), $\dim M<\infty$ and $\mathcal{F}$ is a locally constant sheaf. Then there is a canonical isomorphism

$$H_p^\Phi(M,\mathcal{F})\cong H^{-p}(\Gamma_\Phi(M,\mathcal{D}(\mathcal{J}^\bullet)\otimes\mathcal{F})) \qquad (*)$$

for every $p\geq 0$.

The proof of Theorem 12.10 is easy. If Φ is paracompactifying then $\Delta^k\otimes\mathcal{F}$ and $\mathcal{D}(\mathcal{J}^\bullet)^l\otimes\mathcal{F}$ are Φ-soft (by II.4.8 and II.4.22). If $\mathcal{F}$ is locally constant, then $\Delta^k\otimes\mathcal{F}$ and $\mathcal{D}(\mathcal{J}^\bullet)^l\otimes\mathcal{F}$ are sa-flabby (by II.4.23). In both cases the sheaves $\Delta^k\otimes\mathcal{F}$ and $\mathcal{D}(\mathcal{J}^\bullet)^l\otimes\mathcal{F}$ are Φ-acyclic (II.4.9, II.4.10), and hence

$$\mathrm{H}^{-p}_\Phi(M,\Delta^\bullet\otimes\mathcal{F})=H^{-p}(\Gamma_\Phi(M,\Delta^\bullet\otimes\mathcal{F}))=H_p^\Phi(M,\mathcal{F}).$$
$$\mathrm{H}^{-p}_\Phi(M,\mathcal{D}(\mathcal{J}^\bullet)\otimes\mathcal{F})=H^{-p}(\Gamma_\Phi(M,\mathcal{D}(\mathcal{J}^\bullet)\otimes\mathcal{F})).$$

Remark 12.11. If we defined the semialgebraic Borel-Moore-homology groups by the equality (*), then this definition would be analogue of the definition of the Borel-Moore-homology groups of a locally compact topological space given by Borel and Moore (cf. [BM], [B, V.3]).

Now we assume that M is of type (L) and $\dim M<\infty$. By Theorem 12.10 the homology $H_*(M,\Lambda)$ is just the cohomology of the complex

$$\mathrm{Hom}\,(\Gamma_c(M, \mathcal{J}^\bullet), \Lambda^\bullet).$$

From homological algebra (cf. e.g. [CE, XVII.5.2]) we obtain the exact sequence

(12.12) $$0 \to \mathrm{Ext}\,(H_c^{q+1}(M, \Lambda), \Lambda) \to H_q(M, \Lambda) \to \mathrm{Hom}\,(H_c^q(M, \Lambda), \Lambda) \to 0$$

for every $q \geq 0$.

Finally we give a nice application of the exact sequences (12.12). Suppose M is of type (L) and $\dim M < \infty$. We want to compare the homology of M and M_{loc}. Recall that M_{loc} is defined to be the inductive limit of the system $\gamma_c(M)$ of complete semialgebraic subsets of M (cf. [DK$_3$, I, §7]). Let $p\colon M_{loc} \to M$ be the canonical locally semialgebraic map (loc. cit., as map of sets it is the identity). Let $\Lambda_M \to \mathcal{J}^\bullet$ and $\Lambda_{M_{loc}} \to \mathcal{L}^\bullet$ be injective resolutions and $p^*\mathcal{J}^\bullet \to \mathcal{L}^\bullet$ be a map of resolutions. This map induces a map

$$\Gamma_c(U, \mathcal{J}^\bullet) \longrightarrow \Gamma_c(p^{-1}(U), \mathcal{L}^\bullet)$$

and hence homomorphisms

$$p^* : H_c^*(M, \Lambda) \longrightarrow H_c^*(M_{loc}, \Lambda)$$

and

$$p_* : H_*(M_{loc}, \Lambda) \longrightarrow H_*(M, \Lambda)$$

(although p is in general not a proper map!). By use of p^* and p_* we may compare the exact sequences (12.12) of M and M_{loc}. Since M_{loc} has the same complete semialgebraic subsets as M, the map $p^* : H_c^*(M, \Lambda) \to H_c^*(M_{loc}, \Lambda)$ is clearly an isomorphism. By the Five Lemma we obtain

Theorem 12.13. The canonical map

$$p_* : H_*(M_{loc}, \Lambda) \longrightarrow H_*(M, \Lambda)$$

is an isomorphism.

CHAPTER IV: Some intersection theory

We will discuss some aspects of intersection theory over an algebraically closed field of characteristic zero. More or less we proceed in the same way as Borel and Haefliger ([BH]) in the complex-analytic case.

§1 - Fundamental classes of C^1-manifolds and locally isoalgebraic spaces

Let $U \in \mathring{\gamma}(R^n)$ and $f: U \to R$ be a semialgebraic function. Then f is said to be a C^1-function if, for every $x \in U$, the limits $\frac{\partial f}{\partial x_i}(x) = \lim_{t\to 0} \frac{f(x+te_i)-f(x)}{t}$ exist (where $e_i := (0,\ldots,0,1,0,\ldots,0)$) and $x \mapsto \frac{\partial f}{\partial x_i}(x)$ is a continuous and hence semialgebraic function on U for $i \in \{1,\ldots,n\}$ (cf. [Br, 8.13], [M]). As in differential topology one now defines C^r-functions, C^r-diffeomorphisms and C^r-manifolds (loc. cit.)

Example 1.1. $f: U \to R$ is a C^∞-function if and only if f is a Nash function on U (cf. [Br, 8.13]).

Now let $U, V \in \mathring{\gamma}(R^n)$ and $f: U \to V$ be a C^1-diffeomorphism. If $x \in U$, then we denote the differential of f in x by $\mathrm{df}(x)$. It is the linear automorphism $R^n \to R^n$ given by the matrix $\left(\frac{\partial f_i}{\partial x_j}(x)\right)_{1\le i,j\le n}$. We assume that U (and hence V) is *connected*. Then the restriction maps $H_n(R^n, \mathbf{Z}) \xrightarrow{\text{res}} H_n(U,\mathbf{Z})$ and $H_n(R^n,\mathbf{Z}) \xrightarrow{\text{res}} H_n(V,\mathbf{Z})$ are isomorphisms (III.3.4.c). We want to prove

Proposition 1.2. Let $x \in U$ and $\varepsilon = \operatorname{sign}(\det \mathrm{df}(x))$. Then the diagram

$$\begin{array}{ccc} & H_n(R^n,\mathbf{Z}) & \\ \text{res} \swarrow & & \searrow \text{res} \\ H_n(U,\mathbf{Z}) & \xrightarrow{\varepsilon\cdot f_*} & H_n(V,\mathbf{Z}) \end{array}$$

is commutative.

The proof is divided in several steps.

Lemma 1.3. Let $a \in R^n$ and $T: R^n \to R^n$ be the translation $x \mapsto x + a$. Then

$$T_*: H_*(R^n) \to H_*(R^n)$$

is the identity map.

Proof. $(x,t) \mapsto x + ta \quad (x \in R^n, t \in [0,1])$ is a proper homotopy from id_{R^n} to T. Now apply Theorem III.2.10.

By Lemma 1.3 we may assume from now on in the proof of Prop. 1.2 that $x = f(x) = 0$.

Lemma 1.4. Let A be a linear automorphism of R^n and $\varepsilon = \det A$. Then

$$A_* = \varepsilon \cdot \mathrm{id} : H_*(R^n,\mathbf{Z}) \to H_*(R^n,\mathbf{Z}).$$

Proof. Recall that $H_k(R^n, \mathbf{Z}) = 0$ for $k \neq n$ and $H_n(R^n, \mathbf{Z}) \cong \mathbf{Z}$. Let $e_1, \ldots, e_n$ be the canonical basis of R^n. It is easy to see that the linear automorphism given by $e_1 \mapsto -e_1, e_k \mapsto e_k$ for $k > 1$, induces $-\mathrm{id} : H_*(R^n) \to H_*(R^n)$.

So it suffices to consider the case $\det A > 0$. We know by transfer from the case $R = \mathbf{R}$ that $\mathrm{GL}(n, R)^+ = \{B \in \mathrm{GL}(n, R) \mid \det B > 0\}$ is connected. Therefore we can choose a semialgebraic path $\alpha : [0,1] \to \mathrm{GL}(n, R)^+$ with $\alpha(0) = \mathrm{id}_{R^n}$ and $\alpha(1) = A$. Then

$$H : R^n \times [0,1] \to R^n, \quad (x,t) \mapsto \alpha(t) \cdot x,$$

is a proper homotopy from id_{R^n} to A. Hence $A_* = \mathrm{id} : H_*(R^n) \to H_*(R^n)$ by (III.2.10). Lemma 1.4 is proven.

Proposition 1.2 now follows from

Lemma 1.5. The diagram

$$\begin{array}{ccc} H_*(R^n, \mathbf{Z}) & \xrightarrow{(\mathrm{df}(0))_*} & H_*(R^n, \mathbf{Z}) \\ \mathrm{res} \downarrow & & \downarrow \mathrm{res} \\ H_*(U, \mathbf{Z}) & \xrightarrow{f_*} & H_*(V, \mathbf{Z}) \end{array}$$

commutes.

Proof. Let $\mathcal{H}_n = \mathcal{H}_n(R^n, \mathbf{Z})$ and $x = 0$. It suffices to prove that

$$\begin{array}{ccc} H_*(R^n, \mathbf{Z}) & \xrightarrow{(\mathrm{df}(0))_*} & H_*(R^n, \mathbf{Z}) \\ \downarrow & & \downarrow \\ (\mathcal{H}_n)_x & \xrightarrow{f_*} & (\mathcal{H}_n)_x \end{array}$$

commutes. It is easy to see that $(\mathcal{H}_n)_x = H_n^c(U_1, U_1 \setminus \{x\}, \mathbf{Z})$ for every semialgebraic neighbourhood U_1 of x in R^n where $H_n^c(-,-,\mathbf{Z})$ is the homology with complete supports introduced for arbitrary pairs of affine spaces in [DK$_1$, §3]. Since homotopic maps between pairs of spaces induce the same homomorphisms in $H_*^c(-,-;\mathbf{Z})$ (loc. cit.), it suffices to prove that there is some open semialgebraic neighbourhood U_1 of 0 in U such that $f \mid U_1$ and $\mathrm{df}(0) \mid U_1$ are homotopic as maps of pairs $(U_1, U_1 \setminus \{0\}) \to (R^n, R^n \setminus \{0\})$.

Since $\mathrm{df}(0)$ is a linear automorphism of R^n, there is some $D \in R, D > 0$, such that $\|\mathrm{df}(0)(x)\| \geq D \cdot \|x\|$ for every $x \in R^n$. The Taylor formula for C^1-functions is valid over any real closed field (use Tarski's principle). Hence there is an open semialgebraic neighbourhood U_1 of 0 in U and a constant C such that $0 < C < D$ and $\|f(x) - \mathrm{df}(0)(x)\| \leq C \cdot \|x\|$ for every $x \in U_1$. We conclude that the maps

$$f \mid U_1, \mathrm{df}(0) \mid U_1 : (U_1, U_1 \setminus \{0\}) \to (R^n, R^n \setminus \{0\})$$

are linearly homotopic. The proofs of Lemma 1.5 and Prop. 1.2 are now finished.

Let V be a vector space over R of dimension n. An *orientation* of V is determined by a basis of R. Two bases define the same orientation if, passing from one basis to the other, the matrix of transition has a positive determinant.

If A and B are subspaces of V which are oriented by $(a_1, \dots, a_r)$ and $(b_1, \dots, b_s)$, then the sum $A + B$ is oriented by $(a_1, \dots, a_r, b_1, \dots b_s)$ provided $A \cap B = \emptyset$.
If $V = A \oplus B$ and A and V are oriented, then B will be oriented in such a way that V is the oriented sum of A and B.
Finally if $V = A + B$ and V, A, B are oriented, then $D = A \cap B$ is oriented in the following way: We choose A' and B' with $V = A \oplus A'$ and $V = B \oplus B'$. Then we orient A' and B' as explained above. Now observe that $V = D \oplus (A' \oplus B')$ and we endow D with the complementary orientation of $A' \oplus B'$ in V.

Let $C = R(\sqrt{-1})$ be the algebraic closure of R and W be a vector space over C of (complex) dimension n. Considered as an R-vector space of dimension $2n$, it bears a *natural* orientation: If $e_1, \dots, e_n$ is a basis of W over C, the R-basis $(e_1, \sqrt{-1} \cdot e_1, \dots, e_n, \sqrt{-1} \cdot e_n)$ defines the natural orientation of W.
We assume that R^n is always endowed with its standard orientation $(e_1, \dots, e_n)$ where $e_i = (0, \dots, 0, \overset{i}{1}, 0, \dots, 0)$.

Now we consider an oriented R-vector space V of dimension n. It is a semialgebraic manifold of dimension n over R and its orientation defines a (homological) orientation of this manifold in the sense of III, §3. We explain this in the following. First we consider the case $V = R^n$. Let $c \in H_1(R, \mathbf{Z})$ be the canonical generator of $H_1(R, \mathbf{Z})$. If R is considered as the open 1-simplex $]-\infty, +\infty[$ with vertices $-\infty, +\infty$, then $c = < -\infty, +\infty >$ (cf. III, §1). From Poincaré duality (III, §9) we know that

$$H^n_{\{0\}}(R^n, \mathbf{Z}) \to H^{\{0\}}_0(R^n, \mathbf{Z}), z \mapsto \eta \cap z,$$

is an isomorphism if η is one of the generators of $H_n(R^n, \mathbf{Z}) \cong \mathbf{Z}$. The homology group $H^{\{0\}}_0(R^n, \mathbf{Z}) = H_0(\{0\}, \mathbf{Z})$ has the canonical generator $< 0 >$.

Now let c^* be the generator of $H^1_{\{0\}}(R, \mathbf{Z})$ with $c \cap c^* = < 0 >$.

By p_k we denote the projection $R^n \to R, (x_1, \dots, x_n) \mapsto x_k$. Let $A_k := \{x \in R^n \mid x_k = 0\}$ and $B_r = A_1 \cap \dots \cap A_r$. Note that $z_r := (p_1^* c^*) \cup \dots \cup (p_r^* c^*)$ is an element of $H^r_{B_r}(R^n, \mathbf{Z})$.

Lemma 1.6. i) $p_k^* : H^1_{\{0\}}(R, \mathbf{Z}) \to H^1_{A_k}(R^n, \mathbf{Z})$ is an isomorphism.
ii) z_r is a generator of $H^r_{B_r}(R^n, \mathbf{Z})$ for $1 \le r \le n$.

Proof. i) follows easily from the „homotopy theorem" II.2.5.
ii) is proved by induction on r. The case $r = 1$ follows from i). So suppose that $r > 1$ and z_{r-1} generates $H^{r-1}_{B_{r-1}}(R^n, \mathbf{Z})$. We choose some generator η of $H_n(R^n, \mathbf{Z})$. The inclusion $R^{n-r+1} = \{(0, \dots, 0)\} \times R^{n-r+1} \hookrightarrow R^n$ is denoted by i.

Now we compute by III.8.1.iii that

$$\eta \cap z_r = (\eta \cap z_{r-1}) \cap p_r^* c^*.$$

From Poincaré duality (III, §9) we conclude that $\eta \cap z_{r-1}$ generates $H^{B_{r-1}}_{n-r+1}(R^n, \mathbf{Z})$. Hence there is some generator γ of $H_{n-r+1}(R^{n-r+1}, \mathbf{Z})$ which is mapped to $\eta \cap z_{r-1}$ by the natural isomorphism $i_* : H_{n-r+1}(R^{n-r+1}, \mathbf{Z}) \xrightarrow{\sim} H^{B_{r-1}}_{n-r+1}(R^n, \mathbf{Z})$.
Using Prop. III.8.2 we compute

$$\eta \cap z_r = i_*\gamma \cap p_r^* c^* = i_*(\gamma \cap i^* p_r^* c^*).$$

By i) $i^* p_r^* c^* = (p_r \circ i)^* c^*$ generates $H^1_{\{0\}\times R^{n-r}}(R^{n-r+1}, \mathbf{Z})$ and hence $\gamma \cap i^* p_r^* c$ generates $H^{\{0\}\times R^{n-r}}_{n-r}(R^{n-r+1}, \mathbf{Z})$. This implies that $\eta \cap z_r$ generates $H^{B_r}_{n-r}(R^n, \mathbf{Z})$ and, by Poincaré duality, this means that z_r generates $H^r_{B_r}(R^n, \mathbf{Z})$. q.e.d.

In particular $c_n^* := z_n$ is a generator of $H^n_{\{0\}}(R^n, \mathbf{Z})$. Now we define c_n to be the generator of $H_n(R^n, \mathbf{Z})$ with $c_n \cap c_n^* = < 0 >$. The element c_n is called the *fundamental class* of R^n (associated to the standard orientation of R^n).

If V is an oriented R-vectorspace of dimension n, then we choose a linear isomorphism $h: R^n \xrightarrow{\sim} V$ preserving the orientations. The element $h_*(c_n) \in H_n(V, \mathbf{Z})$ is called the *fundamental class* of V.

Now we consider an n-dimensional locally semialgebraic C^1-manifold M. It posesses an *atlas* $\{(U_i, \varphi_i), i \in I\}$ *of class* C^1, i.e. $(U_i \mid i \in I)$ is an admissible covering of M by open semialgebraic subsets U_i, each $\varphi_i : U_i \xrightarrow{\sim} V_i$ is a semialgebraic isomorphism from U_i onto some open semialgebraic subset V_i of R^n and the map

$$\varphi_j \circ \varphi_i^{-1} : \varphi_i(U_i \cap U_j) \to \varphi_j(U_i \cap U_j)$$

is a C^1-isomorphism for every pair of indices i, j.

Definition. The atlas is said to be *oriented* if $\det d(\varphi_j \circ \varphi_i^{-1})(x) > 0$ for every pair i, j of indices and every $x \in \varphi_i(U_i \cap U_j)$.

Now we assume that the given atlas is oriented. This oriented atlas gives us a homological $\mathbf{Z}$-orientation of M in the sense of chap. III, §3, as follows.

The restriction $c_n \mid V_i$ of the fundamental class of R^n is a fundamental class of V_i, i.e. $H_n(V', \mathbf{Z}) = \mathbf{Z} \cdot (c_n \mid V') \cong \mathbf{Z}$ for every connected component V' of V_i. Hence

$$[U_i] := (\varphi_i^{-1})_*(c_n \mid V_i) \in H_n(U_i, \mathbf{Z})$$

is a fundamental class of U_i. By Proposition 1.2 these fundamental classes $[U_i], i \in I$, glue together to form a fundamental class $[M] \in H_n(M, \mathbf{Z})$ of M.

Example 1.7. Let $C = R(\sqrt{-1})$ be the algebraic closure of R and X be an n-dimensional locally isoalgebraic manifold over C (cf. [Hu], [K]).
Let $\{(U_i, \varphi_i), i \in I\}$ be an isoalgebraic atlas of X, i.e. $(U_i \mid i \in I)$ is an admissible covering of X by open semialgebraic subsets, each φ_i is a semialgebraic isomorphism $U_i \xrightarrow{\sim} V_i$ from U_i onto some open semialgebraic subset V_i of C^n and the transition maps

$$\varphi_j \circ \varphi_i^{-1} : \varphi_i(U_i \cap U_j) \to \varphi_j(U_i \cap U_j)$$

are isoalgebraic isomorphisms ([K, §2]).

We identify C^n with R^{2n} in the usual way: $C \oplus \ldots \oplus C = R \oplus R \cdot \sqrt{-1} \oplus R \oplus \ldots \oplus R \cdot \sqrt{-1}$. Notice that the fundamental class $c_{2n} \in H_{2n}(R^{2n}, \mathbf{Z})$ belongs to the natural orientation of the C-vector space C^n.

Isoalgebraic functions are C^1 (they are even C^∞) and therefore X is a $2n$-dimensional C^1-manifold over R. Since the differentials of an isoalgebraic isomorphism $U \xrightarrow{\sim} V$ ($U, V \in \mathring{\gamma}(C^n)$) always have a positive determinant (same proof as for holomorphic isomorphisms), our given isoalgebraic atlas is oriented and thus yields a fundamental class $[X] \in H_{2n}(X, \mathbb{Z})$. It does not depend on the chosen isoalgebraic atlas. It is called the *canonical fundamental class* of X and determines the *canonical orientation* of X.

At the end of this section we prove that every reduced locally isoalgebraic space $(X, \mathcal{O}_X)$ over $C = R(\sqrt{-1})$ has a canonical fundamental class. We shortly recall the definition and some properties (cf. [Hu]). A *locally isoalgebraic space* $(X, \mathcal{O}_X)$ is a ringed space over R (in the sense of [DK$_3$, §1]) with $\mathcal{O}_X$ a sheaf of C-algebras. It posesses an admissible covering $(X_i \mid i \in I)$ by open subsets such that, for every $i \in I$, the open subspace $(X_i, \mathcal{O}_X \mid X_i)$ is isomorphic to some open isoalgebraic subspace $(U_i, \mathcal{O}_{U_i})$ of some affine scheme V_i over C. This means that U_i is an open semialgebraic subset of $V_i(C)$, the set of C-rational points of V_i, and $\mathcal{O}_{U_i}$ is the sheaf of isoalgebraic functions on U_i (cf. [K, §2]). Of course $V_i(C)$ is here considered as a semialgebraic space over R (via the identification $C = R^2$). The stalks $\mathcal{O}_{X,x}, x \in X$, are local rings. If all these stalks are reduced local rings, then the space $(X, \mathcal{O}_X)$ is called *reduced*. In this case the sets U_i consist of reduced points of the scheme V_i, i.e. we may choose V_i as a (reduced) variety over C.
The sheaves of semialgebraic functions on the sets U_i glue together to form a sheaf $\tilde{\mathcal{O}}_X$ on X. Then $(X, \tilde{\mathcal{O}}_X)$ is a locally complete locally semialgebraic space, called the underlying locally semialgebraic space of X. By $H_*(X, \mathcal{F})$ we mean the homology groups of the underlying locally semialgebraic space.
From now on we assume that all occuring locally isoalgebraic spaces are reduced and that the underlying locally semialgebraic space is paracompact. The set of regular points x of a locally isoalgebraic space X (regular means: $\mathcal{O}_{X,x}$ is a regular local ring) denoted by X_{reg} is a dense open locally semialgebraic subset of X.
A subset Y of X is called *locally isoalgebraic* if there exists an admissible covering $(V_j \mid j \in J)$ of X by open semialgebraic subsets such that $Y \cap V_j$ is the set of zeros of finitely many isoalgebraic functions $f_{jk} \in \Gamma(V_j, \mathcal{O}_X)$. Every locally isoalgebraic subset Y of X may be endowed with the reduced subspace structure $(Y, \mathcal{O}_Y)$, and in this way it also becomes a reduced locally isoalgebraic space. $X \setminus X_{\text{reg}}$ is a locally isoalgebraic subset of X.
As in complex analysis the irreducible components of $(X, \mathcal{O}_X)$ are the closures (in the strong topology) of the connected components of X_{reg}. They are locally isoalgebraic subsets and form a locally finite (cf. [DK$_3$, I. §1, Def. 4]) family of subsets of X. Every irreducible component has a well defined complex dimension. X is said to be of pure dimension n if every irreducible component of X has dimension n. Then X_{reg} is an n-dimensional isoalgebraic manifold over C. In general $\dim X$ is defined to be the supremum of the dimensions of its irreducible components. The dimension of the underlying locally semialgebraic space over R is denoted by $\dim_R X$. Notice that $\dim_R X = 2 \cdot \dim X$.

Example 1.8. Let V be a (reduced) algebraic variety over C. Then $(V, \mathcal{O}_V)$ where $\mathcal{O}_V$ is the sheaf of isoalgebraic functions on V (cf. [K]) is a reduced isoalgebraic space over C.

The structure sheaf is usually omitted in our notation.

Theorem 1.9. Let X be a reduced locally isoalgebraic space over C. Assume that X is of pure dimension n. Then there is a uniquely determined class $[X] \in H_{2n}(X, \mathbb{Z})$ called the *canonical fundamental class* of X such that $[X] \mid X_{reg}$ is the canonical fundamental class of the n-dimensional isoalgebraic manifold X_{reg}. If X is irreducible then $H_{2n}(X, \mathbb{Z}) = \mathbb{Z} \cdot [X] \cong \mathbb{Z}$.

The proof is almost trivial. The locally isoalgebraic subset $X \setminus X_{reg}$ of X has at most the (semialgebraic) dimension $2n-2$ (over R). Hence we conclude from the long exact sequence III.2.8 that $H_{2n}(X, \mathbb{Z}) \xrightarrow{\text{res}} H_{2n}(X_{reg}, \mathbb{Z})$ is an isomorphism. If X is irreducible, then X_{reg} is connected and the second statement follows.

§2 - Intersection of homology classes

Let M be a paracompact Λ-oriented semialgebraic manifold over R and $[M] \in H_n(M,\Lambda)$ be its fundamental class (cf. III, §3).
The cap product with $[M]$ yields an isomorphism

$$\begin{aligned} H^p_\Phi(M,\Lambda) &\xrightarrow{\sim} H^\Phi_{n-p}(M,\Lambda) \\ \alpha &\longmapsto [M]\cap\alpha \end{aligned}$$

for every p and every family Φ of supports on M (Poincaré duality, III, §9). We always denote the inverse map

$$H^\Phi_{n-p}(M,\Lambda) \longrightarrow H^p_\varphi(M,\Lambda)$$

by the letter δ.

Let Φ, Ψ be families of supports on M. Then we may define an intersection product

$$\begin{aligned} H^\Phi_i(M,\Lambda) \times H^\Psi_j(M,\Lambda) &\longrightarrow H^{\Phi\cap\Psi}_{i+j-n}(M,\Lambda) \\ (a,b) &\longmapsto a\cdot b \end{aligned}$$

by

$$a\cdot b := \delta^{-1}(\delta a \cup \delta b).$$

Here $\cup$ denotes the cup product

$$H^{n-i}_\Phi(M,\Lambda)\otimes H^{n-j}_\Psi(M,\Lambda) \longrightarrow H^{2n-(i+j)}_{\Phi\cap\Psi}(M,\Lambda)$$

(cf. [B, II.7]).

Proposition 2.1. Let $a \in H^\Phi_i(M,\Lambda)$ and $b \in H^\Psi_j(M,\Lambda)$.

i) The intersection product is bilinear.

ii) We have $a\cdot b = a \cap \delta b$

iii) The intersection product is associative, i.e. if Θ is another support family on M and $c \in H^\Theta_k(M,\Lambda)$, then

$$(a\cdot b)\cdot c = a\cdot(b\cdot c) \in H^{\Phi\cap\Psi\cap\Theta}_{i+j+k-2n}(M,\Lambda).$$

iv) $a\cdot b = (-1)^{(n-i)(n-j)} b\cdot a$.

v) If $a' \in \Gamma(M,\Delta_i)$ and $b' \in \Gamma(M,\Delta_j)$ are representatives of a and b such that $\dim(\operatorname{supp}(a')\cap\operatorname{supp}(b')) < i+j-n$, then $a\cdot b = 0$.

Proof. i) is trivial.

ii) $a\cdot b = \delta^{-1}(\delta a\cup\delta b) = [M]\cap(\delta a\cup\delta b) = ([M]\cap\delta a)\cap\delta b = a\cap\delta b$.

iii) $(a\cdot b)\cdot c = (a\cap\delta b)\cap\delta c = a\cap(\delta b\cup\delta c) = a\cap\delta(b\cdot c) = a\cdot(b\cdot c)$.

iv) This is true since $\delta a\cup\delta b = (-1)^{(n-i)\cdot(n-j)}\delta b\cup\delta a$ (cf. [B, II.7]).

v) Let $| a' | = \operatorname{supp}(a')$ and $| b' | = \operatorname{supp}(b')$. Our claim follows from the commutative diagram

$$
\begin{array}{ccc}
H_i^{|a'|}(M,\Lambda) \times H_j^{|b'|}(M,\Lambda) & \stackrel{\cdot}{\longrightarrow} & H_{i+j-n}^{|a'|\cap|b'|}(M,\Lambda) = 0 \\
\downarrow & & \downarrow \\
H_i^{\Phi}(M,\Lambda) \times H_j^{\Psi}(M,\Lambda) & \stackrel{\cdot}{\longrightarrow} & H_{i+j-n}^{\Phi\cap\psi}(M,\Lambda).
\end{array}
$$

Let V be an n-dimensional vector space over R and W be an oriented subvector space of dimension r. The inclusion map $W \hookrightarrow V$ is denoted by i_W, the fundamental class of W in $H_r(W,\mathbf{Z})$ by w (cf. §1). By $(i_W)_*$ we denote the canonical isomorphism $H_*(W,\mathbf{Z}) \xrightarrow{\sim} H_*^W(V,\mathbf{Z})$. The projection $V \to W$ is denoted by p_W.
Let w^* be the dual fundamental class of w, i.e. the uniquely determined element $w^* \in H^r_{\{0\}}(W,\mathbf{Z})$ with $w \cap w^* =< 0 >\in H_0^{\{0\}}(W,\mathbf{Z})$. If V is the oriented sum $W \oplus W'$ of the oriented subvector spaces W and W', then $v^* = p_W^*(w^*) \cup p_{W'}^*(w'^*)$ (cf. §1).

Proposition 2.2. Let A and B be oriented subvector spaces of dimension r and s of the n-dimensional oriented R-vector space V. Suppose $V = A + B$. Then the following statements hold.

a) If $A \cap B = \emptyset$ and V is the oriented sum $A + B$ of A and B, then we have

$$\delta(i_B)_* b = p_A^* a^*, \quad \delta(i_A)_* a = (-1)^{rs} p_B^* b^*$$

and hence

$$i_A^*(\delta(i_B)_* b) = a^*, \quad i_B^*(\delta(i_A)_* a) = (-1)^{rs} b^*.$$

b) If $W = A \cap B$ is oriented as described in §1, then we have

$$((i_A)_* a) \cdot ((i_B)_* b) = (i_W)_* w.$$

The proof can be copied word by word from [BH, 2.10] and therefore it is omitted.

We give an application of Prop. 2.2 which will be needed in the next section.

Proposition 2.3. Let V and W be locally isoalgebraic manifolds of (complex) dimension n and m over the algebraic closure $C = R(\sqrt{-1})$ of R. Let $X \subset V$ be a locally isoalgebraic subset of V. Assume that X is of pure dimension r. Let $p: V \times W \to V$ be the projection and $j: X \hookrightarrow V$ and $i: X \times W \hookrightarrow V \times W$ be the inclusion maps. Then we have

$$p^*(\delta j_*[X]) = \delta i_*[X \times W]$$

in $H^{2n-2r}_{X\times W}(V \times W, \mathbf{Z})$.

Recall that $[X] \in H_{2r}(X,\mathbf{Z})$ and $[X \times W] \in H_{2m+2r}(X \times W, \mathbf{Z})$ are the canonical fundamental classes of X and $X \times W$ (cf. §1).

In the proof of Prop. 2.3 we need the following

Lemma 2.4. Let M be an oriented n-dimensional manifold over R and $[M] \in H_n(M,\mathbf{Z})$ be its fundamental class. Let N be a *connected* r-dimensional semialgebraic submanifold of M and $U \in \mathring{\gamma}(M)$ with $U \cap N \neq \emptyset$. Then the canonical map

$$H_N^{n-r}(M,\mathbf{Z}) \longrightarrow H_{N\cap U}^{n-r}(U,\mathbf{Z})$$

is injective.

Proof. We consider the commutative diagram (coefficients $\mathbf{Z}$)

$$\begin{array}{ccc} H_N^{n-r}(M) & \xrightarrow{\sim} & H_r^N(M) = H_r(N) \\ \downarrow & & \downarrow \text{res} \\ H_{N\cap U}^{n-r}(U) & \xrightarrow{\sim} & H_r^{N\cap U}(U) = H_r(N\cap U) \end{array}$$

where the horizontal arrows are the Poincaré duality isomorphisms. We either have $H_r(N)=0$ or $H_r(N)\cong \mathbf{Z}$. In both cases the restriction map $H_r(N)\to H_r(N\cap U)$ is injective since N is connected and $U\cap N\neq\emptyset$. Our claim follows.

Proof of Proposition 2.3. Replacing V by $V\setminus(X\setminus X_{\text{reg}})$ we may assume that X is an isoalgebraic submanifold of dimension r. We may also assume that all spaces V,W,X are connected. Let $U\in\mathring{\gamma}(N)$ and $U'\in\mathring{\gamma}(W)$ such that $U\cap X\neq\emptyset$ and $U'\neq\emptyset$. By the preceding Lemma 2.4 it suffices to prove that

$$p^*(\delta j_*[X])\mid U\times U' = \delta i_*[X\times W]\mid U\times U'.$$

Choosing for U and U' suitable coordinate neighbourhoods we may therefore assume that $V=C^n, W=C^m$ and $X=\{0\}\times C^r\subset C^n$. Let $Y=C^{n-r}\times\{0\}\subset C^n$. Now we compute by use of Prop. 2.2 a)

$$\begin{aligned} &p^*(\delta j_*[X]) = p^*(p_{V,Y}^*[Y]^*) = (p_{V,Y}\circ p)^*([Y]^*) = \\ &p_{V\times W,Y}^*([Y]^*) = \delta i_*[X\times W]. \end{aligned}$$

Here $p_{V,Y}$ is the projection $V\to Y$ and $p_{V\times W,Y}$ the projection $V\times W\to Y$. Proposition 2.2.a) was used in the first and in the last equality. (Note that $V=Y\oplus X$ and $V\times W= Y\oplus(X\oplus W)$).

§3 - Intersection of isoalgebraic cycles

Let C be an algebraically closed field of characteristic zero. We choose a real closed field $R \subset C$ with $C = R(\sqrt{-1})$ and consider a paracompact locally isoalgebraic manifold V of dimension n over C (e.g. V a smooth algebraic variety). V is oriented in a canonical way (cf. Example 1.7). We will define the intersection cycle of two locally isoalgebraic cycles on V.

Let $(X_i)_{i \in I}$ be the set of irreducible locally isoalgebraic subsets of V.

Definition 1. The group $\mathfrak{Z}(V)$ of *locally isoalgebraic cycles* consists of the formal sums $Z = \Sigma n_i \cdot X_i$ with coefficients in $\mathbf{Z}$ such that the sets X_i with $n_i \neq 0$ form a locally finite family of V. The sets X_i with $n_i \neq 0$ are called the *components* of Z. The *support* $\mid Z \mid$ of Z is defined to be the union of the components of Z. A cycle Z is pure of dimension p if every component of Z has dimension p.

If X is a locally isoalgebraic subset of V and $X_\lambda, \lambda \in J$, are its irreducible components, then we often denote the cycle $\sum_{\lambda \in J} X_\lambda$ simply by X.

Every p-dimensional irreducible locally isoalgebraic subset X of V determines a generator $h(X)$ of $H^X_{2p}(V, \mathbf{Z})$, namely the image of the fundamental class $[X]$ under the canonical isomorphism $H_{2p}(X, \mathbf{Z}) \xrightarrow{\sim} H^X_{2p}(V, \mathbf{Z})$ (cf. Theorem 1.9). If $Z = \Sigma n_i \cdot X_i \in \mathfrak{Z}(V)$, then $h(Z) \in H^{|Z|}_*(V, \mathbf{Z})$ is defined to be the sum $\Sigma n_i \cdot h(X_i)$ in the sense of III.2.9. If Φ is a family of supports on V containing $\mid Z \mid$, then we denote the image of $h(Z)$ in $H^\Phi_*(V, \mathbf{Z})$ by $h^\Phi(Z)$. For example, $h^V(Z)$ is the image of $h(Z)$ in $H_*(V, \mathbf{Z})$. The application

$$Z \longmapsto h^V(Z)$$

is an additive homomorphism from $\mathfrak{Z}(V)$ to $H_*(V, \mathbf{Z})$.

Lemma 3.1. Let X be a (reduced) paracompact locally isoalgebraic space of pure dimension p and $(X_\lambda \mid \lambda \in I)$ be its irreducible components. Then the canonical map

$$\prod_{\lambda \in I} H_{2p}(X_\lambda, \mathbf{Z}) \longrightarrow H_{2p}(X, \mathbf{Z})$$

$$(a_\lambda \mid \lambda \in I) \longmapsto \Sigma a_\lambda$$

(cf. III.2.9) is an isomorphism.

The easy proof may be copied from [BH, 4.3]. The reason why Lemma 3.1 is true is that $X_\lambda \cap X_\mu$ has at most dimension $2p - 2$ over R if $\lambda \neq \mu$.

Now we set out to define intersection multiplicities. Let $X, Y \in \mathfrak{Z}(V)$ be of pure dimension and D be an irreducible component of $X \cap Y$.

Definition 2. D is said to be *proper* if $\operatorname{codim} D = \operatorname{codim} X + \operatorname{codim} Y$ (codim = codimension in V).

Let D_λ $(\lambda \in I)$ be the proper components of $X \cap Y$ and M be their union and $p = \dim D_\lambda$. By N we denote the union of the non proper components. Since $\dim_R M \cap N \leq 2p - 2$, we conclude from (III.2.11) that

$$H_{2p}^{|X|\cap|Y|}(V,\mathbf{Z}) = H_{2p}^{M}(V,\mathbf{Z}) \oplus H_{2p}^{N}(V,\mathbf{Z}).$$

By Lemma 3.1 we obtain a canonical isomorphism

$$H_{2p}^{|X|\cap|Y|}(V,\mathbf{Z}) = \prod_{\lambda\in I} H_{2p}^{D_\lambda}(V,\mathbf{Z}) \oplus H_{2p}^{N}(V,\mathbf{Z}).$$

Every element $c \in H_{2p}^{|X|\cap|Y|}(V,\mathbf{Z})$ can be (uniquely) written as a sum

$$c = \sum_{\lambda\in I} c_\lambda + d, \quad c_\lambda \in H_{2p}^{D_\lambda}(V,\mathbf{Z}), d \in H_{2p}^{N}(V,\mathbf{Z})).$$

Let $h(X)\cdot h(Y) \in H_{2p}^{|X|\cap|Y|}(V,\mathbf{Z})$ be the intersection of the homology classes $h(X)$ and $h(Y)$ on the (canonically) oriented manifold V.

Definition 3. The component of $h(X)\cdot h(Y)$ in $H_{2p}^{D_\lambda}(V,\mathbf{Z}) = \mathbf{Z}\cdot h(D_\lambda)$ is a multiple $m\cdot h(D_\lambda)$. The integer m is denoted by $i(X\cdot Y, D_\lambda)$ and called the *intersection multiplicity* of the proper component D_λ in $X\cap Y$.

Proposition 3.2. Let $U \in \mathring{\gamma}(V)$ be given with $U\cap D_\lambda \neq \emptyset$. Let $D_{\lambda,U}$ be an irreducible component of $D_\lambda \cap U$. Then we have

$$i(X\cdot Y, D_\lambda) = i((X\cap U)\cdot(Y\cap U), D_{\lambda,U}).$$

This is obvious from the definition since $h(X\cap U) = h(X)\mid U$ and the intersection product in homology is compatible with restriction.

Definition 4. Two irreducible locally isoalgebraic subsets X and Y of V *intersect properly* if every component D of $X\cap Y$ is proper.

Now let $X, Y \in \mathfrak{Z}(V)$ be cycles such that every component of X intersects every component of Y properly. Then we may define the intersection product $X\cdot Y$ as follows:

If X and Y are irreducible and D_λ, $\lambda\in I$, are the irreducible components of $X\cap Y$, then we set

$$X\cdot Y := \sum_{\lambda\in I} i(X\cdot Y, D_\lambda)\cdot D_\lambda.$$

In general, if $X = \Sigma n_i X_i$ and $Y = \Sigma n_j Y_j$, then we define

$$X\cdot Y := \Sigma n_i\cdot m_j(X_i\cdot Y_j).$$

Notice that this sum is well defined and locally finite since the sums we started with are locally finite.

Now we will gather some properties of the intersection product. It is associative.

Proposition 3.3. Let $X, Y, Z \in \mathfrak{Z}(V)$ be pure dimensional and D be a proper component of $X \cap Y \cap Z$ (i.e. codim $D =$ codim $X +$ codim $Y +$ codim Z). Let $D_\lambda, \lambda \in I$, (resp. $E_\mu, \mu \in J$) be the irreducible components of $X \cap Y$ (resp. $Y \cap Z$) containing D. Then each D_λ (resp. E_μ) is proper in $X \cap Y$ (resp. $Y \cap Z$) and we have

$$\sum_\lambda i(X \cdot Y, D_\lambda) \cdot i(D_\lambda \cdot Z, D) = \sum_\mu i(X \cdot E_\mu, D) \cdot i(Y \cdot Z, E_\mu).$$

Proof. The same proof as in [BH, 4.7] applies. It is based on the associativity of the intersection product in homology (Prop. 2.1.iii)).

Proposition 3.4. Let X and Y be pure dimensional locally isoalgebraic subsets of V and D be a proper component of $X \cap Y$. Let $x \in D$ be a point which is regular on X and Y. Suppose that X and Y intersect transversally in x (i.e. the tangent space T_xV of V in x is generated by the tangent spaces T_xX and T_xY). Then

$$i(X \cdot Y, D) = 1.$$

Proof. By Prop. 3.2 we may replace V by any open semialgebraic neighbourhood U of x Choosing U small enough we have an isoalgebraic isomorphism $\Phi: U \xrightarrow{\sim} V$ onto some open semialgebraic subset V of C^n such that $\Phi(x) = 0, \Phi(X \cap U) = A \cap V$ and $\Phi(Y \cap U) = B \cap V$ where $A = C^r \times \{0\} \subset C^n$ and $B = \{0\} \times C^s \subset C^n$. (Use the implicit function theorem for isoalgebraic functions). Then

$$i(X \cdot Y, D) = i((A \cap V) \cdot (B \cap V), A \cap B \cap V).$$

Again applying Prop. 3.2 we see that

$$i(X \cdot Y, D) = i(A \cdot B, A \cap B).$$

From Prop. 2.2.b) we conclude that $i(A \cdot B, A \cap B) = 1$.

Let W be another paracompact locally isoalgebraic manifold over C. Let $X \subset V$ and $Y \subset V \times W$ be pure dimensional locally isoalgebraic subsets and assume that the image $p(Y)$ under the projection $p: V \times W \to V$ is a locally isoalgebraic subset of V. Suppose there is some open locally semialgebraic subset U of $p(Y)$ with the following properties:

a) $A := p(Y) \setminus U$ is a locally isoalgebraic subset of V.
b) $\dim A < \dim p(Y)$
c) $U \cap X \neq \emptyset$
d) $p \mid p^{-1}(U) \cap Y : p^{-1}(U) \cap Y \to U$ is an isoalgebraic isomorphism.

Now we consider an irreducible locally isoalgebraic subset D of Y such that $D \cap p^{-1}(U) \neq \emptyset$ and $p(D)$ is a locally isoalgebraic subset of V. In this situation we have

Proposition 3.5. (Projection formula) D is a proper component of $(X \times W) \cap Y$ if and only if $p(D)$ is a proper component of $X \cap p(Y)$, and in this case we have

$$i(X \cdot p(Y), p(D)) = i((X \times W) \cdot Y, D).$$

Proof. Note that $\dim p(Y) = \dim Y$ and $\dim p(D) = \dim D$. Moreover, $p(D)$ is irreducible. Since $\operatorname{codim}_{V\times W}(X \times W) + \operatorname{codim}_{V\times W}(Y) = \operatorname{codim}_V(X) + \operatorname{codim}_V(p(Y)) + \dim W$ and $\operatorname{codim}_{V\times W}(D) = \operatorname{codim}_V(p(D)) + \dim W$, the first statement is obvious.
To prove the equality we may assume by Prop. 3.2 (replacing V by $V \setminus A$) that $Y \to p(Y)$ is an isomorphism. From Prop. 2.3 we know that $\delta h(X \times W) = p^*(\delta h(X))$. Now we conclude by use of (III.8.2) and Prop. 2.1 that

$$h(p(Y)) \cdot h(X) = (p_* h(Y)) \cdot h(X) = (p_* h(Y)) \cap \delta h(X) =$$
$$p_*(h(Y) \cap p^*(\delta h(X))) = p_*(h(Y) \cap \delta h(X \times W)) =$$
$$p_*(h(Y) \cdot h(X \times W)).$$

Since $Y \xrightarrow{\sim} p(Y)$ and hence $h(p(D)) = p_*(h(D))$, the equality follows from the definition of the intersection multiplicities. q.e.d.

Now we consider an isoalgebraic subset X of some open semialgebraic neighbourhood $U' \in \mathring{\gamma}(C^n)$ of $0 \in C^n$ which is pure of dimension p. Let L be an $(n-p)$–dimensional C-linear subspace of C^n such that 0 is an isolated point of $X \cap L$. Replacing U' by a smaller set, we assume that $X \cap L \cap U' = \{0\}$. Applying a suitable linear coordinate transformation we may assume that $L = \{0\} \times C^{n-p} \subset C^n$.
By $\mathcal{O}_{C^p,0}$ and $\mathcal{O}_{C^n,0}$ we denote the stalks of the sheaves of isoalgebraic functions on C^p and C^n in 0. Let $p: C^n \to C^p, (z_1, \dots, z_n) \mapsto (z_1, \dots, z_p)$ be the projection. Let $\mathfrak{a} \subset \mathcal{O}_{C^n,0}$ be the ideal consisting of those germs of isoalgebraic functions which vanish on the germ of X in 0. There are finitely many minimal prime ideals $\mathfrak{p}_1, \dots, \mathfrak{p}_r$ over $\mathfrak{A}$ in $\mathcal{O}_{C^n,0}$. Let $K = \operatorname{Quot}(\mathcal{O}_{C^p,0})$ and $K_i = \operatorname{Quot}(\mathcal{O}_{C^n,0/\mathfrak{p}_i}), 1 \le i \le r$, be the quotient fields and $\mu = \sum_{i=1}^{r} [K_i : K]$ be the sum of the degrees of the field extensions $K \subset K_i$.
Then there exist open polycylindres $U_1 \subset C^p$ and $U_2 \subset C^{n-p}$ containing 0 and a proper isoalgebraic subset Δ of U_1 such that the following holds (with $U := U_1 \times U_2$), cf. [Hu, §3, §5]:

i) The restriction $\pi := p \mid X \cap U : X \cap U \longrightarrow U_1$ is proper and semialgebraic.

ii) Every $z \in U_1 \setminus \Delta$ has precisely μ preimages $Y_1, \dots, Y_\mu$ in $X \cap U$ under π.

iii) For every $z \in U_1 \setminus \Delta$ the hyperplane $z + L = p^{-1}(z)$ intersects X in all points of $X \cap U \cap p^{-1}(z)$ transversally.

Proposition 3.6. $i(X \cdot L, 0) = \mu$.

Proof. We choose an injective semialgebraic path $\gamma: [0,1] \to U_1$ with $\gamma(0) = 0$ and $\gamma(]0,1]) \subset U_1 \setminus \Delta$. Let $L_t := \gamma(t) + L = p^{-1}(\gamma(t))$. Then we have

a) $L_0 = L$.

b) L_t intersects X in U in μ points and the intersection in each of these points is transversal for $t > 0$.

c) $X \cap U \cap \pi^{-1}(\gamma[0,1]) = X \cap U \cap (\bigcup_t L_t)$ is a complete semialgebraic set (by i).

It suffices to show that $i(X \cdot L, 0) = \# L_1 \cap X \cap U$. This may be done by the same arguments as those in [BH, 4.10].

Corollary 3.7. Suppose X is an algebraic (reduced) subvariety of C^n. Then $i(X \cdot L, 0)$ coincides with the algebraic intersection multiplicity ν of X and L in 0.

Proof. We have to show $\mu = \nu$. Let $C[X]$ be the coordinate ring of X and $\mathcal{O}_{X,0} = \mathcal{O}_{C^n,0/a}$ be the local ring of germs of isoalgebraic functions on X in 0.

By definition ν is the multiplicity $e(\mathfrak{q})$ of the ideal $\mathfrak{q}$ of $C[X]_{(0)}$ generated by the coordinate functions $Z_1, \ldots, Z_p$ (cf. [Sa, II, §5]) and $e(\mathfrak{q})$ is equal to the multiplicity $e(\mathfrak{q}')$ of the ideal $\mathfrak{q}'$ of $C\hat{[X]}_{(0)}$ generated by $Z_1, \ldots, Z_p$ (cf. [ZS, Lemma 1 on p. 285]). By ^ we denote the completion). By [ZS, Cor. 2 on p. 300] $e(\mathfrak{q}')$ is equal to $[C\hat{[X]}_{(0)} : C\hat{[Z_1, \ldots, Z_p]}_{(0)}]$. (Here we denote by $[B:A]$ the maximal number of elements of B which are linearly independent over A). Since $\mathcal{O}_{X,0}$ and $\mathcal{O}_{C^p,0}$ are the henselizations of $C[X]_{(0)}$ and $C[Z_1, \ldots, Z_p]_{(0)}$, we have $C\hat{[X]}_{(0)} = \hat{\mathcal{O}}_{X,0}$ and $\hat{\mathcal{O}}_{C^p,0} = C\hat{[Z_1, \ldots, Z_p]}_{(0)}$. Thus ν is equal to $[\hat{\mathcal{O}}_{X,0} : \hat{\mathcal{O}}_{C^n,0}]$ and this number in turn is equal to $[\mathcal{O}_{X,0} : \mathcal{O}_{C^n,0}]$ (cf. [Ab, p. 243]). But $[\mathcal{O}_{X,0} : \mathcal{O}_{C^n,0}]$ is just the number μ defined above (cf. [Ab, 9.25]). Thus indeed $\mu = \nu$.

Proposition 3.8. Suppose V is a smooth algebraic variety over C and X, Y are irreducible subvarieties of V. Let D be a proper component of $X \cap Y$. Then $i(X \cdot Y, D)$ coincides with the algebraic intersection multiplicity of X and Y in D (cf. e.g. [Sa], [Se] for the definition).

Proof. In view of Prop. 3.3, Prop. 3.4, Prop. 3.5 and Cor. 3.7 this follows from [W, Appendix II].

Remark 3.9. Consider the situation above. Proposition 3.8 says in particular that we would get the same result if we computed the intersection multiplicity $i(X \cdot Y, D)$ with respect to another real closed field $R' \subset C$ with $C = R'(\sqrt{-1})$.

Again let V be a paracompact locally isoalgebraic manifold over C.

Definition 5. Two locally isoalgebraic cycles Z_1, Z_2 on V are said to be *isoalgebraically equivalent* if there exists a connected locally isoalgebraic manifold W, a locally isoalgebraic cycle Z in $V \times W$ and points $w_1, w_2 \in W$ such that $(V \times w_1) \cdot Z$ and $(V \times w_2) \cdot Z$ are defined and coincide with Z_1 and Z_2 (after identifying $V \times w_1$ and $V \times w_2$ with V).

Proposition 3.10. Let Z_1, Z_2 be two cycles on V which are isoalgebraically equivalent. Then $h^V(Z_1) = h^V(Z_2)$.

Proof. Let $p: V \times W \to V$ and $q: V \times W \to W$ be the projections and Φ be the family of closed locally semialgebraic subsets A of $V \times W$ such that $p \mid A$ is proper. The projection q induces a homomorphism $q^*: H_c^*(W, \mathbf{Z}) \to H_\Phi^*(V \times W, \mathbf{Z})$. The points w_1 and w_2 define the same class $[w] \in H_c^0(W, \mathbf{Z})$ since W is connected. Now we derive from Prop. 2.3 that $\delta h^\Phi(V \times w_1) = q^*(\delta[w]) = \delta h^\Phi(V \times w_2)$. Hence $h^\Phi(V \times w_1) = h^\Phi(V \times w_2)$ and hence $h^V(Z_1) = p_*(h^\Phi(V \times w_1) \cdot h^{V \times W}(Z)) = p_*(h^\Phi(V \times w_2) \cdot h^{V \times W}(Z)) = h^V(Z_2)$.

Remark 3.11. Let V be a quasiprojective smooth algebraic variety over C. Then Prop. 3. and the preceding Prop. 3.10 imply that $Z \mapsto h^V(Z)$ induces a ring homomorphism

$$h_*: A_*(V) \longrightarrow H_*(V, \mathbf{Z})$$

from the Chow ring $A_*(V)$ of V to the homology of V with closed supports and coefficients in $\mathbb{Z}$. (Of course the multiplication on $H_*(V,\mathbb{Z})$ is given by the intersection product defined in §2).

Remark 3.12. Fulton developed an algebraic intersection theory on arbitrary (not necessarily smooth and quasiprojective) varieties V over C (cf. [F]). In Chapter 19 of his book he compares the algebraic and topological intersection theory in the case $C = \mathbb{C}$. In particular he studies a cycle map $A_*(V) \longrightarrow H_*(V,\mathbb{Z})$ from the Chow ring to topological Borel-Moore-homology which generalizes the homomorphism h_* in Remark 3.11. Since now semialgebraic Borel-Moore-homology is available over arbitrary real closed fields and has all nice properties of topological Borel-Moore-homology (cf. [F, 19.1]), it is possible to generalize many of Fulton's results in [F, Chap. 19] to arbitrary algebraically closed groundfields C of characteristic 0.

LIST OF SYMBOLS

INDEX

REFERENCES

[Ab] **S. Abhyankar**, Local analytic geometry, Academic Press, New York and London, 1964.

[A] **M. Artin**, Grothendieck topologies, Harvard University, 1962.

[BH] **A. Borel, A. Haefliger**, La classe d'homologie fondamentale d'un espace analytique, Bull. soc. math. France 89 (1961), 461 - 513.

[BM] **A. Borel, J.C. Moore**, Homology theory for locally compact spaces, Mich. math. J. 7 (1960), 137 - 159.

[BCR] **J. Bochnak, M. Coste, M.-F. Roy**, Géométrie algébrique réelle, Ergebnisse der Mathematik und ihrer Grenzgebiete, 3. Folge, Band 12, Springer-Verlag 1987.

[B] **G. Bredon**, Sheaf theory, McGraw-Hill, 1967.

[Brö] **L. Bröcker**, Real spectra and distributions of signatures, in: Géométrie algébrique réelle et formes quadratiques, Proceedings, Rennes 1981, Springer Lecture Notes in Mathematics 959, 1982.

[Br] **G.W. Brumfiel**, Partially ordered rings and semialgebraic geometry, Cambridge University Press, 1979.

[Br_1] **G.W. Brumfiel**, Witt rings and K-theory, Rocky Mountain J. Math. 14 (4) (1984), 733 - 765.

[CE] **H. Cartan, S. Eilenberg**, Homological algebra, Princeton University Press, 1956.

[CC] **M. Carral, M. Coste**, Normal spectral spaces and their dimensions, J. Pure Appl. Algebra 301 (1983), 227 - 235.

[CR] **M. Coste, M.F. Roy**, La topologie du spectre réel, Contemporary Mathematics, Vol. 8, Ordered fields and real algebraic geometry, 27 - 59, Amer. Math. Soc., Providence, 1982.

[D] **H. Delfs**, Kohomologie affiner semialgebraischer Räume, Diss., Regensburg, 1980.

[D_1] **H. Delfs**, Semialgebraic Borel-Moore-homology, Rocky Mountain J. Math. 14 (1984), 987 - 990.

[D_2] **H. Delfs**, The homotopy axiom in semialgebraic cohomology, J. reine angew. Math.355 (1985), 108 - 128.

[DK] **H. Delfs, M. Knebusch**, Semialgebraic topology over a real closed field II: Basic theory of semialgebraic spaces, Math. Z. 178 (1981), 175 - 213.

[DK_1] **H. Delfs, M. Knebusch**, On the homology of algebraic varieties over real closed fields, J. reine angew. Math. 335 (1981), 122 - 163.

[DK$_2$] **H. Delfs, M. Knebusch**, Separation, retractions and homotopy extension in semialgebraic spaces, Pacific J. Math. 114 (1984), 47 - 71.

[DK$_3$] **H. Delfs, M. Knebusch**, Locally semialgebraic spaces, Lecture Notes in Mathematics 1173, Springer-Verlag 1985.

[DK$_4$] **H. Delfs, M. Knebusch**, An introduction to locally semialgebraic spaces, Rocky Mountain J. Math.14 (1984), 945 - 963.

[Do] **A. Dold**, Lectures on algebraic topology, Springer-Verlag, 1972.

[DV] **A. Douady, J.L. Verdier**, Séminaire de géométrie analytique, Astérisque 36/37, (1976).

[F] **W. Fulton**, Intersection theory, Springer-Verlag 1984.

[G] **R. Godement**, Topologie algébrique et théorie des faisceaux, Hermann, Paris 1964.

[Gr] **A. Grothendieck**, Sur quelques points d'algébre homologique, Tohoku Math. J. 9 (1957), 119 - 221.

[H] **R. Hartshorne**, Residues and duality, Springer Lecture Notes in Mathematics 20, 1966.

[Hu] **R. Huber**, Isoalgebraische Geometrie, Diss., Universität Regensburg, 1984.

[K] **M. Knebusch**, Isoalgebraic geometry: First steps, Séminaire de Théorie des Nombres, Delange-Pisot-Poitou, Paris 1980/81, in: Progess in Mathematics 22, Birkhäuser 1982, 127 - 140.

[K$_1$] **M. Knebusch**, An invitation to real spectra. In: C.R. Riehm, J. Hambleton (eds.), 1983 Conference on Quadratic Forms and Hermitian K-Theory, Canadian Mathematical Society Conference Proceedings 4, 1984.

[K$_2$] **M. Knebusch**, Weakly semialgebraic spaces, Lecture Notes in Mathematics, vol. 1367, Springer-Verlag 1989.

[L] **T.Y. Lam**, An introduction to real algebra, Lecture Notes Sexta Escuela Latinoamericana Matematica, Oaxtepex, Mexico, 1982.

[M] **G. Meixner**, Differentialtopologische Untersuchungen über semialgebraische Mengen und Abbildungen, Universität Regensburg, 1984.

[Ro] **R. Robson**, Embedding semialgebraic spaces, Math. Z. 183 (1983), 365 - 370.

[R] **M.F. Roy**, Spectre réel d'un anneau et topos étale réel, Thèse, Université Paris Nord (1980).

[Sa] **P. Samuel**, Méthodes d'algèbre abstrait en géométrie algébrique, Springer-Verlag 1967.

[Sch] **H. Schubert**, Topologie, Teubner 1964.

[S] **N. Schwartz**, Real closed spaces, Habilitationsschrift, Universität München 1984.

[S_1] **N. Schwartz**, The basic theory of real closed spaces, Memoirs of the AMS, Vol. 77, Number 397, 1989.

SHS] Seminaire Heidelberg-Strasbourg 1966/67, Dualité de Poincaré, Publication I.R.M.A. Strasbourg, 1969.

[Se] **J.P. Serre**, Algèbre locale - multiplicités, Springer Lecture Notes in Mathematics 11 (1965).

[Sw] **R. Swan**, The theory of sheaves, University of Chicago Press 1964.

[V] **J.L. Veridier**, Dualité dans la cohomologie des espaces localement compact, Seminaire Bourbaki, vol. 300 (1965).

[W] **A. Weil**, Foundations of algebraic geometry, Amer. Math. Soc. 1962.

[Wh] **G.W. Whitehead**, Elements of Homotopy Theory, Springer-Verlag, 1978.

[ZS] **O. Zariski, P. Samuel**, Commutative Algebra II, Van Nostrand 1960.